Algebra 2 Workbook

Essential Learning Math Skills
Plus Two Algebra 2 Practice Tests

By

Michael Smith & Reza Nazari

Algebra 2 Workbook

Published in the United State of America By

The Math Notion

Web: WWW.MathNotion.Com

Email: info@MathNotion.com

Copyright © 2020 by the Math Notion. All rights reserved. No part of this publication may be reproduced, stored in a retrieval system, or transmitted in any form or by any means, electronic, mechanical, photocopying, recording, scanning, or otherwise, except as permitted under Section 107 or 108 of the 1976 United States Copyright Ac, without permission of the author.

All inquiries should be addressed to the Math Notion.

About the Author

Michael Smith has been a math instructor for over a decade now. He holds a master's degree in Management. Since 2006, Michael has devoted his time to both teaching and developing exceptional math learning materials. As a Math instructor and test prep expert, Michael has worked with thousands of students. He has used the feedback of his students to develop a unique study program that can be used by students to drastically improve their math score fast and effectively.

- **– SAT Math Practice Book**
- **– ACT Math Practice Book**
- **– PSAT Math Practice Book**
- **– Algebra Math Practice Books**
- **– Common Core Math Practice Books**
- **–many Math Education Workbooks, Exercise Books and Study Guides**

As an experienced Math teacher, Mr. Smith employs a variety of formats to help students achieve their goals: He tutors online and in person, he teaches students in large groups, and he provides training materials and textbooks through his website and through Amazon.

You can contact Michael via email at:

info@Mathnotion.com

Prepare for the Algebra 2 with a Perfect Workbook!

Algebra 2 Workbook is a learning workbook to prevent learning loss. It helps you retain and strengthen your Math skills and provides a strong foundation for success. This Algebra book provides you with solid foundation to get a head starts on your upcoming Algebra Test.

Algebra 2 Workbook is designed by top math instructors to help students prepare for the Algebra course. It provides students with an in-depth focus on the Algebra concepts. This is a prestigious resource for those who need an extra practice to succeed on the Algebra test.

Algebra 2 Workbook contains many exciting and unique features to help you score higher on the Algebra test, including:

- Over 2,500 Algebra Practice questions with answers
- Complete coverage of all Math concepts which students will need to ace the Algebra test.
- Two Algebra 2 practice tests with detailed answers
- Content 100% aligned with the latest Algebra courses.

This Comprehensive Workbook for the Algebra is a perfect resource for those Algebra takers who want to review core content areas, brush-up in math, discover their strengths and weaknesses, and achieve their best scores on the Algebra test.

WWW.MathNotion.COM

… So Much More Online!

- ✓ FREE Math Lessons
- ✓ More Math Learning Books!
- ✓ Mathematics Worksheets
- ✓ Online Math Tutors

For a PDF Version of This Book

Please Visit WWW.MathNotion.com

contents

Chapter 1: Review Linear Functions 11

    Relation and Functions 12

    Finding Slope 13

    Graphing Lines Using Line Equation 14

    Writing Linear Equations 15

    Graphing Linear Inequalities 16

    Write an Equation from a Graph 17

    Rate of change 18

    x and y intercepts 18

    Slope–intercept Form 19

    Point–slope Form 20

    Graphing Lines of Equations 21

    Equation of Parallel or Perpendicular Lines 22

    Equations of Horizontal and Vertical Lines 23

    Graphing Absolute Value Equations 24

    Systems of Equations 25

    Systems of Equations Word Problems 26

    Systems of 3 Variable Equations 27

    Answers of Worksheets – Chapter 1 28

Chapter 2: Monomials and polynomials 35

    GCF of Monomials 36

    Factoring Quadratics 37

    Factoring by Grouping 38

    GCF and Powers of Monomials 39

    Writing Polynomials in Standard Form 40

    Simplifying Polynomials 41

    Adding and Subtracting Polynomials 42

    Multiplying a Polynomial and a Monomial 43

    Multiplying Binomials 44

    Factoring Trinomials 45

Operations with Polynomials 46

Answers of Worksheets – Chapter 6 47

Chapter 3: Radicals Expressions 53

Square Roots 54

Simplifying Radical Expressions 55

Multiplying Radical Expressions 56

Simplifying Radical Expressions Involving Fractions 57

Adding and Subtracting Radical Expressions 58

Answers of Worksheets – Chapter 3 59

Chapter 4: Functions Operations and Quadratic 61

Evaluating Function 62

Adding and Subtracting Functions 63

Multiplying and Dividing Functions 64

Composition of Functions 65

Quadratic Equation 66

Solving Quadratic Equations 67

Quadratic Formula and the Discriminant 68

Quadratic Inequalities 69

Graphing Quadratic Functions 70

Domain and Range of Radical Functions 71

Solving Radical Equations 72

Answers of Worksheets – Chapter 4 73

Chapter 5: Rational Expressions 78

Simplifying and Graphing Rational Expressions 79

Adding and Subtracting Rational Expressions 80

Multiplying and Dividing Rational Expressions 81

Solving Rational Equations and Complex Fractions 82

Answers of Worksheets – Chapter 5 83

Chapter 6: Matrices 85

Adding and Subtracting Matrices 86

Matrix Multiplication 87
Finding Determinants of a Matrix 88
Finding Inverse of a Matrix 89
Matrix Equations 90
Answers of Worksheets – Chapter 6 91

Chapter 7: Sequences and Series 93

Arithmetic Sequences 94
Geometric Sequences 95
Comparing Arithmetic and Geometric Sequences 96
Finite Geometric Series 97
Infinite Geometric Series 98
Answers of Worksheets – Chapter 7 99

Chapter 8: Complex Numbers 103

Adding and Subtracting Complex Numbers 104
Multiplying and Dividing Complex Numbers 105
Graphing Complex Numbers 106
Rationalizing Imaginary Denominators 107
Answers of Worksheets – Chapter 8 108

Chapter 9: Logarithms 109

Rewriting Logarithms 110
Evaluating Logarithms 111
Properties of Logarithms 112
Natural Logarithms 113
Exponential Equations and Logarithms 114
Solving Logarithmic Equations 115
Answers of Worksheets – Chapter 9 116

Chapter 10: Conic Sections 119

Equation of a Parabola 120
Focus, Vertex, and Directrix of a Parabola 121
Standard Form of a Circle 122

Equation of Each Ellipse 123
Hyperbola in Standard Form 124
Conic Sections in Standard Form 125
Answers of Worksheets – Chapter 10 126

Chapter 11: Trigonometric Functions 129

Trig ratios of General Angles 130
Sketch Each Angle in Standard Position 131
Finding Co-terminal Angles and Reference Angles 132
Angles and Angle Measure 133
Evaluating Trigonometric Functions 134
Missing Sides and Angles of a Right Triangle 135
Arc Length and Sector Area 136
Answers of Worksheets – Chapter 11 137

Chapter 12: Statistics and Probability 139

Mean and Median 140
Mode and Range 141
Probability Problems 142
Factorials 143
Combinations and Permutations 144
Answers of Worksheets – Chapter 12 145

Algebra 2 Practice Tests 147

Algebra 2 Practice Test 1 151
Algebra 2 Practice Test 2 159

Answers and Explanations 167

Answer Key 167
Practice Tests 1 169
Practice Tests 2 175

Chapter 1:
Review Linear Functions

Topics that you will practice in this chapter:

- ✓ Relation and Function
- ✓ Finding Slope
- ✓ Graphing Lines Using Line Equation
- ✓ Writing Linear Equations
- ✓ Graphing Linear Inequalities
- ✓ Write an Equation from a Graph
- ✓ Finding Rate of Change, x–intercept and y–intercept
- ✓ Slope-Intercept Form
- ✓ Point-Slope Form
- ✓ Graphing Lines of Equations
- ✓ Equation of parallel or perpendicular lines
- ✓ Equations of horizontal and vertical lines
- ✓ Graphing Absolute Value Equation
- ✓ Systems of Equations
- ✓ Systems of 3 variable Equations

"Sometimes the questions are complicated, and the answers are simple." – Dr. Seuss

Relation and Functions

✎ **State the domain and range of each relation. Then determine whether each relation is a function.**

1)
Function:
....
Domain:
....
Range:
....

2 → 1	
4 → 0	
8 → 5	
9 → 7	
10 → 14	

2)
Function:
....
Domain:
....
Range:
....

x	y
2	3
1	0
−1	−4
7	−4
9	5

3)
Function:
....
Domain:
....
Range:
....

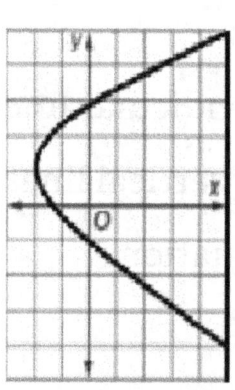

4) $\{(6, -8), (3, -2), (2, 4), (3, 0), (5, 9)\}$
Function:
....
Domain:
....
Range:
....

5)
Function:
....
Domain:
....
Range:
....

6)
Function:
....
Domain:
....
Range:
....

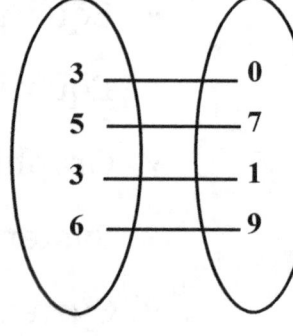

Algebra 2 Workbook

Finding Slope

✎ **Find the slope of each line.**

1) $y = 2x + 5$

2) $y = -x + 17$

3) $y = 4x + 16$

4) $y = -3x + 15$

5) $y = 27 + 7x$

6) $y = 11 - 4x$

7) $y = 7x + 14$

8) $y = -8x + 18$

9) $y = -9x + 15$

10) $y = 8x - 13$

11) $y = \frac{1}{5}x + 9$

12) $y = -\frac{3}{7}x + 19$

13) $-3x + 6y = 17$

14) $4x + 4y = 16$

15) $8y - 3x = 32$

16) $11y - 3x = 42$

✎ **Find the slope of the line through each pair of points.**

17) $(1, 8), (5, 16)$

18) $(-2, 14), (2, 18)$

19) $(7, -1), (3, 9)$

20) $(-4, -4), (2, 14)$

21) $(16, -1), (4, 11)$

22) $(-21, 5), (-10, 38)$

23) $(8, 11), (12, 19)$

24) $(22, -22), (10, 14)$

25) $(21, -15), (19, -13)$

26) $(11, 10), (7, -2)$

27) $(5, 4), (9, 16)$

28) $(34, -87), (22, 45)$

Graphing Lines Using Line Equation

✍ **Sketch the graph of each line.**

1) $y = x - 5$

2) $y = -3x + 4$

3) $x - 2y = 0$

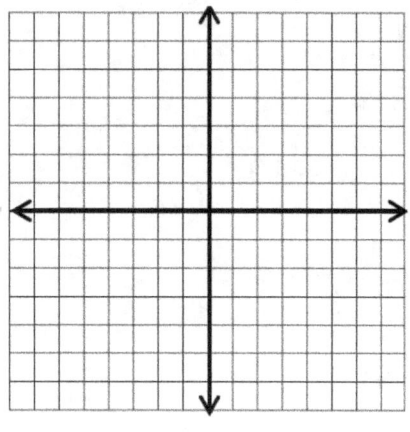
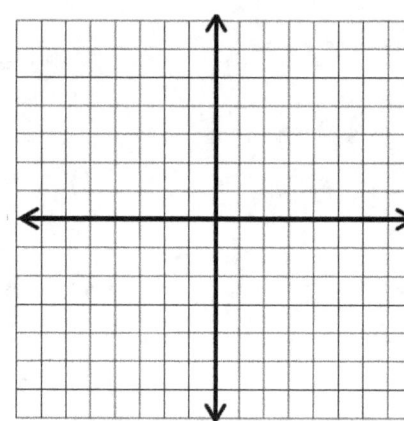
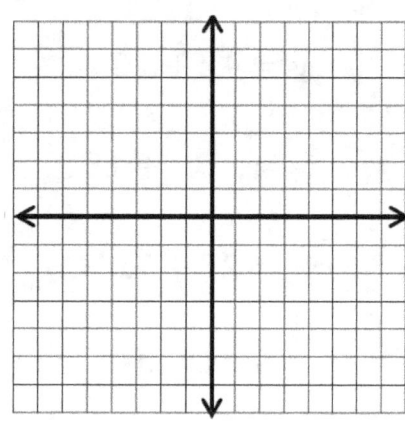

4) $x + y = -4$

5) $4x + 3y = -2$

6) $y - 3x + 6 = 0$

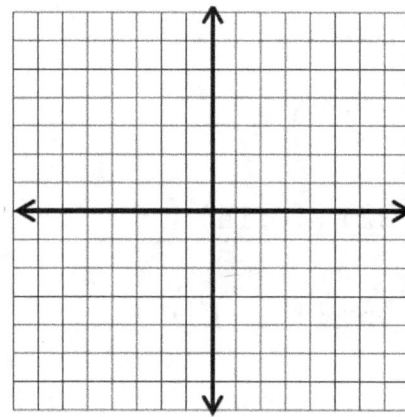
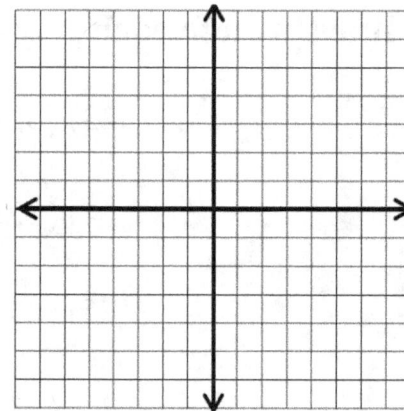
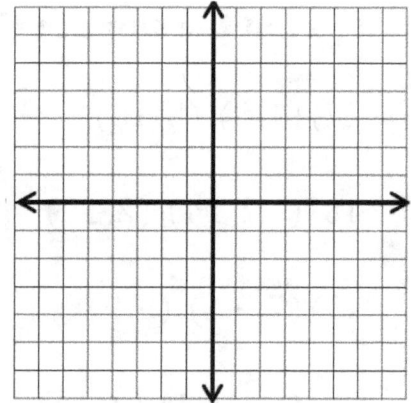

Writing Linear Equations

✎ **Write the equation of the line through the given points.**

1) Through: $(6, -10), (10, 14)$

2) Through: $(10, 4), (4, 22)$

3) Through: $(-6, 4), (2, 12)$

4) Through: $(15, 11), (3, -1)$

5) Through: $(-5, 33), (9, 5)$

6) Through: $(20, 5), (17, 2)$

7) Through: $(24, -4), (16, 4)$

8) Through: $(-18, 57), (33, -45)$

9) Through: $(10, 12), (8, 18)$

10) Through: $(25, 41), (33, -7)$

11) Through: $(-6, 9), (-8, -7)$

12) Through: $(8, 8), (4, -8)$

13) Through: $(6, -10), (10, 6)$

14) Through: $(10, -24), (-8, 12)$

15) Through: $(10, 10), (-2, -4)$

16) Through: $(-7, 35), (11, -31)$

✎ **Find the answer for each problem.**

17) What is the equation of a line with slope 3 and intercept 11? _____

18) What is the equation of a line with slope 5 and intercept 15? _____

19) What is the equation of a line with slope 7 and passes through point $(3, 2)$? _____

20) What is the equation of a line with slope -3 and passes through point $(-2, 5)$? _____

21) The slope of a line is -6 and it passes through point $(-2, 1)$. What is the equation of the line? _____

22) The slope of a line is 5 and it passes through point $(-4, 2)$. What is the equation of the line? _____

Algebra 2 Workbook

Graphing Linear Inequalities

✍ **Sketch the graph of each linear inequality.**

1) $y > 3x - 5$

2) $y < 2x + 1$

3) $y \leq -4x - 5$

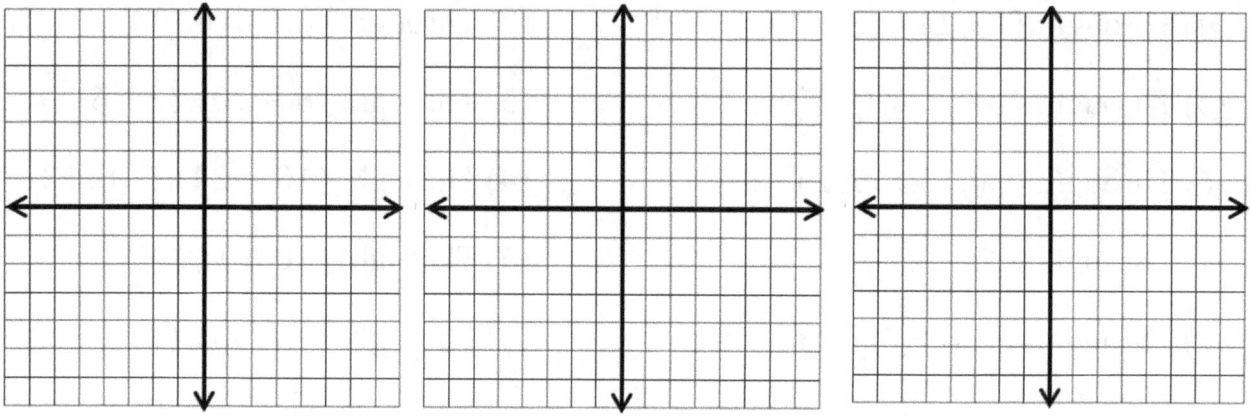

4) $2y \geq 12 + 4x$

5) $-5y < x - 15$

6) $3y \geq -9x + 6$

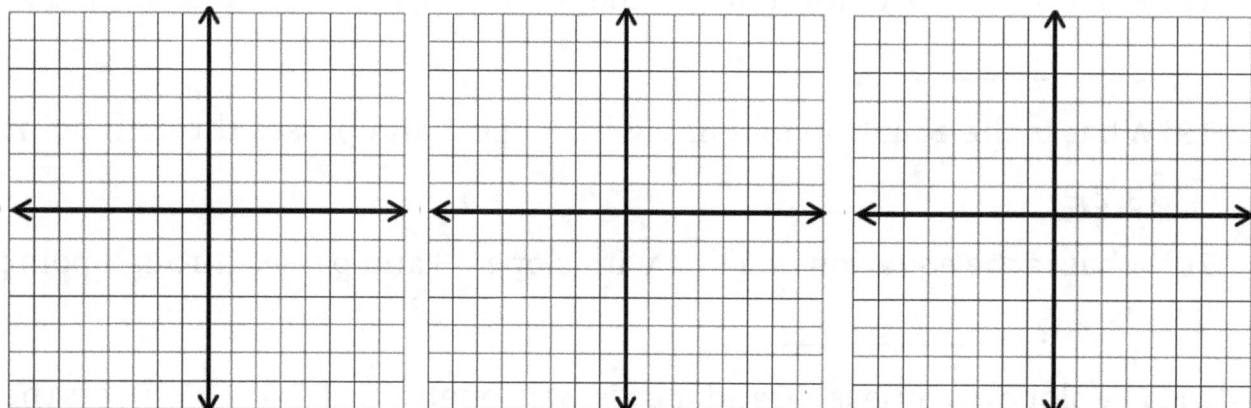

Write an Equation from a Graph

✍ Write the slope intercept form of the equation of each line

1)

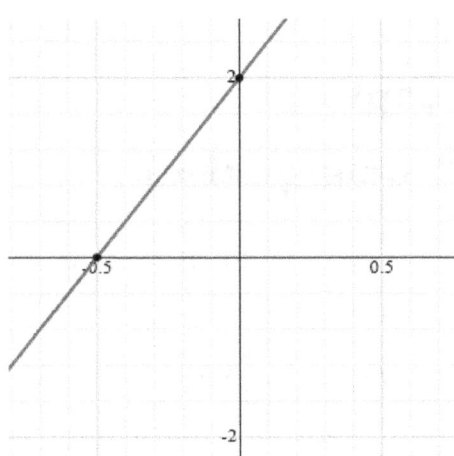

2)

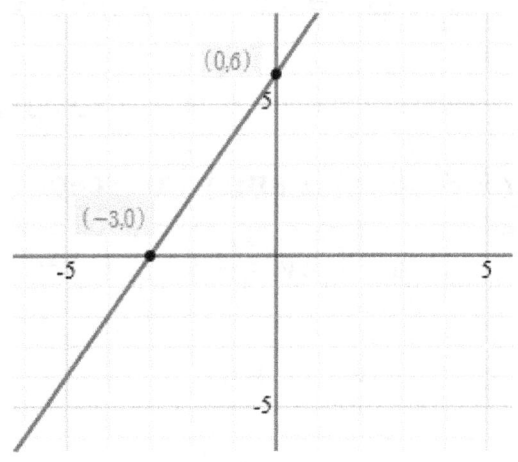

3)

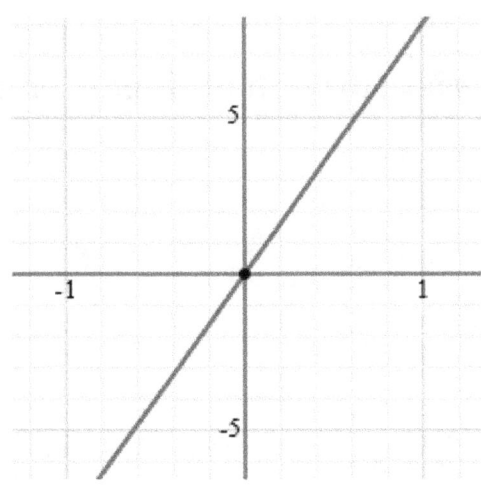

4)

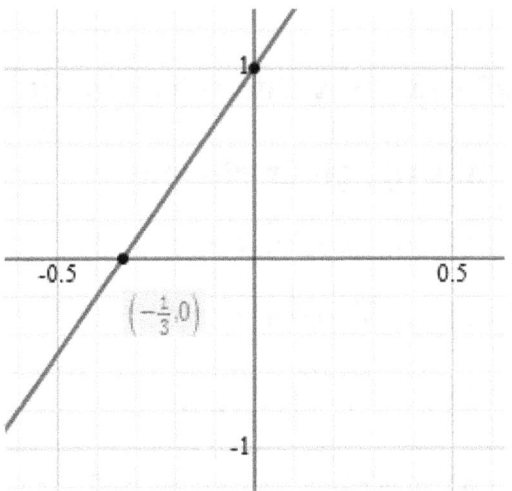

Rate of change

👉 **What is the average rate of change of the function?**

1) $f(x) = 2x^2 + 3$ from $x = 2$ to $x = 5$?

2) $f(x) = -x^2 - 6$, from $x = 3$ to $x = 7$?

3) $f(x) = 2x^3 + 5$, from $x = 0$ to $x = 1$?

x and y intercepts

👉 **Find the x and y intercepts for the following equations.**

1) $4x + 2y = 12$

2) $y = x + 4$

3) $3x = y + 18$

4) $x + y = -6$

5) $7x - 5y = 8$

6) $5y - 4x + 12 = 0$

7) $\frac{3}{5}x + \frac{1}{5}y + \frac{3}{4} = 0$

8) $6x - 24 = 0$

9) $28 - 7y = 0$

10) $-3x - 5y + 45 = 15$

👉 **Find the value of b: The line that passes through each pair of points has the given slope.**

11) $(8, -3), (4, b), m = 2$

12) $(b, 5), (-5, 2), m = \frac{1}{2}$

13) $(-3, b), (3, 5), m = \frac{1}{3}$

14) $(-2, 2), (b, 9), m = 1\frac{3}{4}$

Slope–intercept Form

✏ **Write the slope–intercept form of the equation of each line.**

1) $-15x + y = 7$

2) $-3(5x + y) = 36$

3) $-7x - 21y = -42$

4) $4x + 12 = -8y$

5) $2x - 5y = 15$

6) $14x - 10y = -20$

7) $27x - 9y = -54$

8) $6x - 5y + 36 = 0$

9) $-\frac{1}{4}y = -3x + 5$

10) $8 - 2y - 5x = 0$

11) $-2y = -3x - 8$

12) $12x + 7y = -21$

13) $4(x + 3y + 4) = 0$

14) $y - 6 = 2x + 5$

15) $4(y + 2) = 3(x - 2)$

16) $\frac{2}{5}y + \frac{3}{5}x + \frac{4}{5} = 0$

Point–slope Form

Find the slope of the following lines. Name a point on each line.

1) $y = 3(x + 5)$

2) $y + 2 = \frac{1}{4}(x - 3)$

3) $y + 1 = -2.5x$

4) $y - 4 = \frac{1}{3}(x - 4)$

5) $y + 5 = 0.6(x + 7)$

6) $y - 6 = -2x$

7) $y - 10 = -2(x - 9)$

8) $y + 18 = 0$

9) $y + 19 = 6(x + 1)$

10) $y - 14 = -3(x - 2)$

Write an equation in point–slope form for the line that passes through the given point with the slope provided.

11) $(9, -7), m = 5$

12) $(-2, 5), m = \frac{1}{2}$

13) $(0, -4), m = -3$

14) $(-a, b), m = n$

15) $(-8, 2), m = 4$

16) $(6, 1), m = -4$

17) $(-7, 12), m = \frac{1}{6}$

18) $(0, 13), m = 0$

19) $\left(-\frac{1}{2}, 2\right), m = \frac{1}{7}$

20) $(0, 0), m = -2$

Graphing Lines of Equations

Sketch the graph of each line.

1) $y = 3x - 2$

2) $y = -\frac{1}{2}x + \frac{3}{2}$

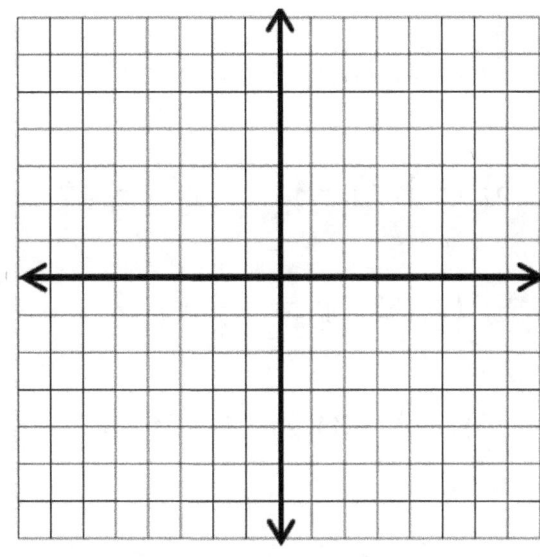
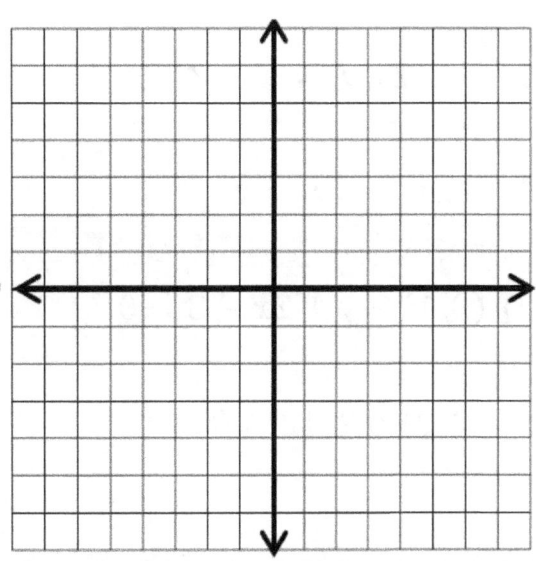

3) $4x - 5y = 12$

4) $-3x - y = 5$

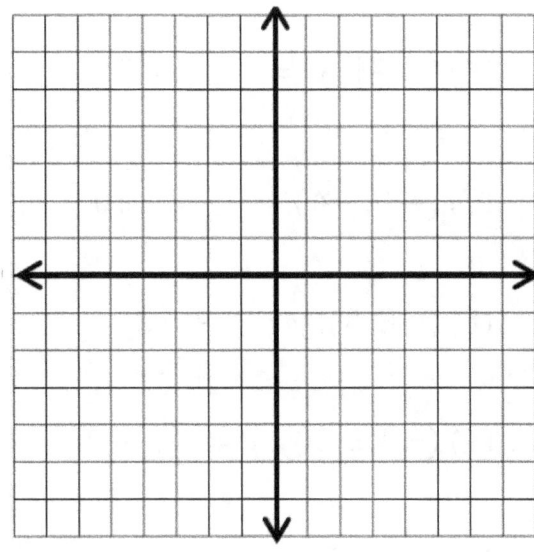
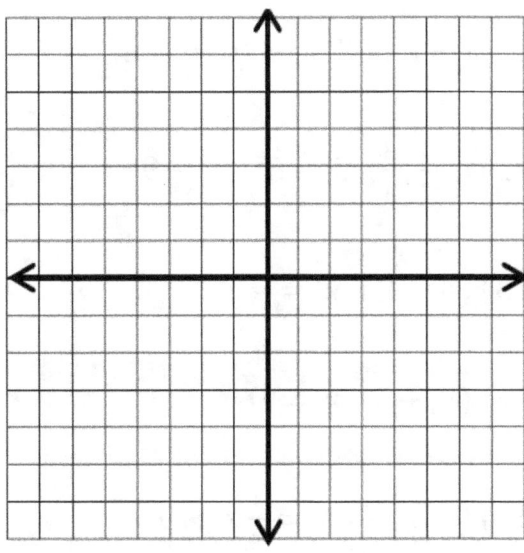

Equation of Parallel or Perpendicular Lines

✏️ Write an equation of the line that passes through the given point and is parallel to the given line.

1) $(-3, -1), x + 2y = -10$

2) $(-3, 2), y = x - 4$

3) $(-3, 1), 4y = x - 7$

4) $(0, 1), -y + 2x - 8 = 0$

5) $(2, 8), y + 12 = 0$

6) $(1, 4), -4x - 2y = -5$

7) $(-3, 0), y = \frac{2}{3}x + 4$

8) $(-1, 3), -4x + y = -16$

9) $(1, -1), y = -\frac{1}{5}x - 2$

10) $(-3, -3), 2x + 10y = -20$

✏️ Write an equation of the line that passes through the given point and is perpendicular to the given line.

11) $(-4, 0), 2x + y = -8$

12) $(-\frac{1}{2}, \frac{3}{4}), 9x - 6y = -9$

13) $(4, -8), y = -8$

14) $(9, -5), x = 9$

15) $(-8, 7), y = \frac{1}{4}x + 9$

16) $(\frac{1}{3}, \frac{2}{3}), y = -4x + 2$

17) $(-8, -4), y = \frac{7}{4}x + 10$

18) $(-8, 5), y = x + 13$

19) $(-4, -10), y = \frac{9}{4}x - 1$

20) $(5, 2), 5y - x + 8 = 13$

Equations of Horizontal and Vertical Lines

✏️ Sketch the graph of each line.

1) $y = -2$

2) $y = 1$

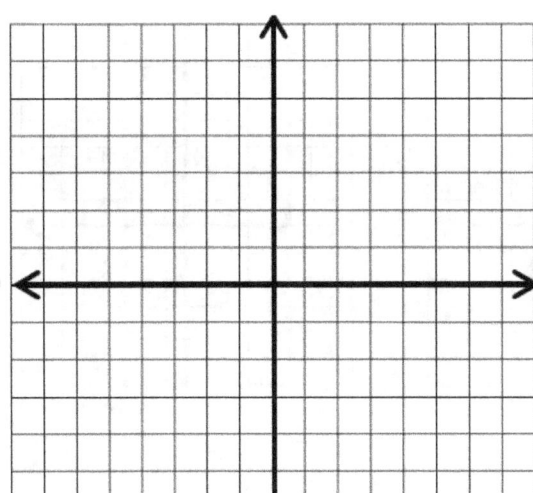
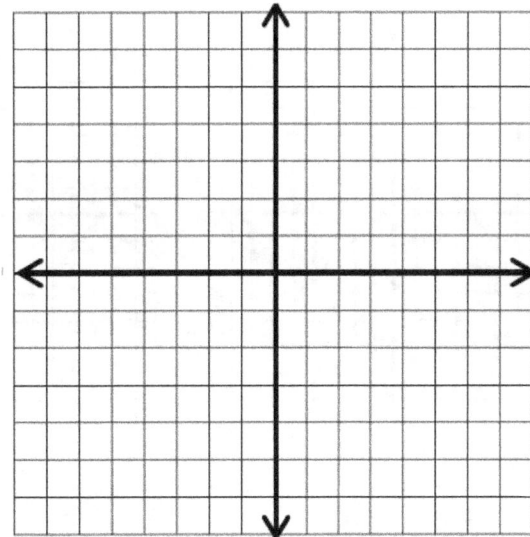

3) $x = -1$

4) $x = 2$

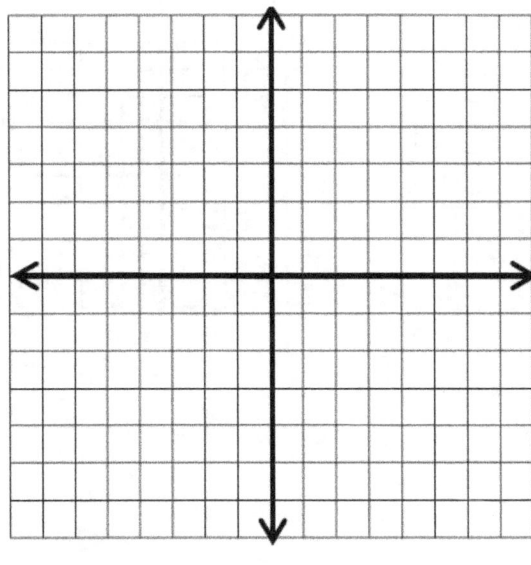
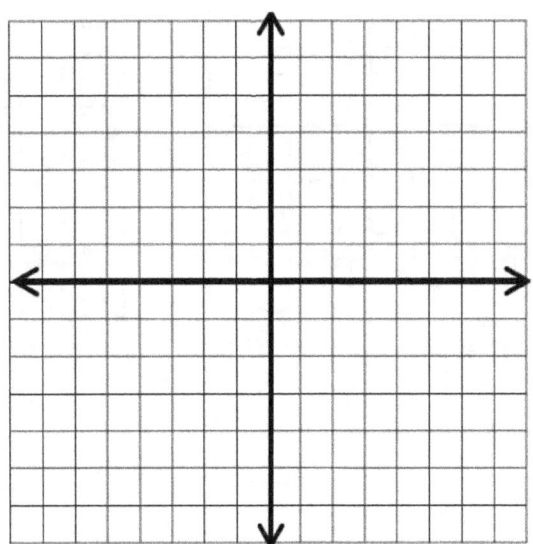

Graphing Absolute Value Equations

✍ **Graph each equation.**

1) $y = |x + 4|$

2) $y = |x + 1|$

3) $y = -|x| - 1$

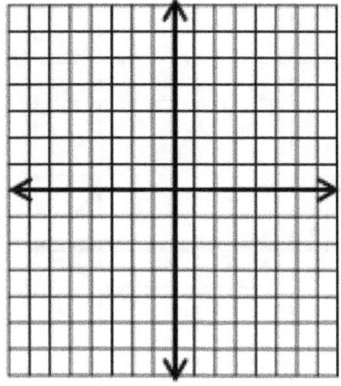 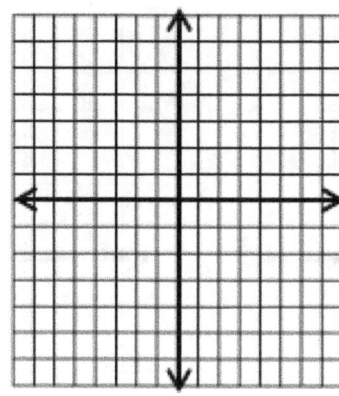 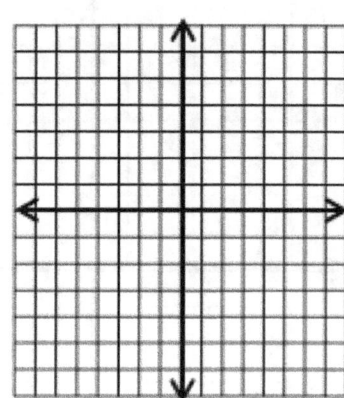

4) $y = |x - 2|$

5) $y = -|x - 2|$

6) $y = -2|2x + 2| + 4$

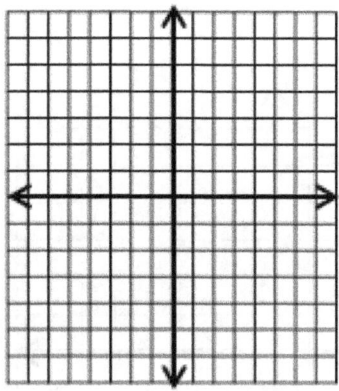 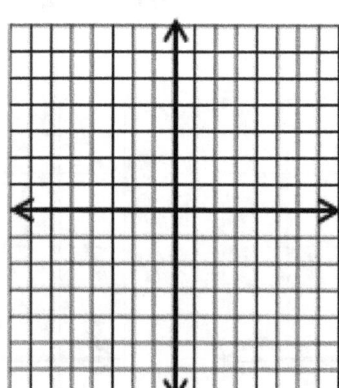 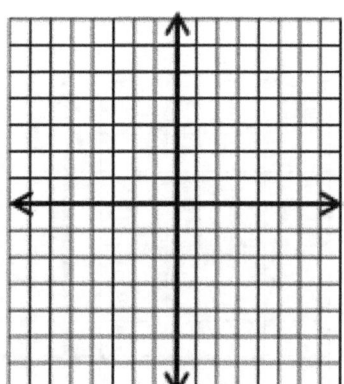

Systems of Equations

Calculate each system of equations.

1) $-6x + 7y = 8$ $x = $ ___
   $x + 4y = 9$ $y = $ ___

2) $-4x + 12y = 12$ $x = $ ___
   $14x - 16y = 10$ $y = $ ___

3) $y = -9$ $x = $ ___
   $2x - 5y = 12$ $y = $ ___

4) $4y = -4x + 20$ $x = $ ___
   $8x - 2y = -12$ $y = $ ___

5) $10x - 9y = -13$ $x = $ ___
   $-5x + 3y = 11$ $y = $ ___

6) $-6x - 8y = 10$ $x = $ ___
   $4x - 8y = 20$ $y = $ ___

7) $5x - 14y = -23$ $x = $ ___
   $-6x + 7y = 8$ $y = $ ___

8) $-4x + 3y = 3$ $x = $ ___
   $-x + 2y = 5$ $y = $ ___

9) $-4x + 5y = 15$ $x = $ ___
   $-3x + 4y = -10$ $y = $ ___

10) $-6x - 6y = -21$ $x = $ ___
    $-6x + 6y = -66$ $y = $ ___

11) $12x - 21y = 6$ $x = $ ___
    $-6x - 3y = -12$ $y = $ ___

12) $-4x - 4y = -14$ $x = $ ___
    $4x - 4y = 44$ $y = $ ___

13) $4x + 5y = 3$ $x = $ ___
    $3x - y = 6$ $y = $ ___

14) $3x - 2y = 2$ $x = $ ___
    $10x - 10y = 20$ $y = $ ___

15) $5x + 8y = 14$ $x = $ ___
    $-3x - 2y = -3$ $y = $ ___

16) $8x + 5y = 4$ $x = $ ___
    $-3x - 4y = 15$ $y = $ ___

Algebra 2 Workbook

Systems of Equations Word Problems

✎ **Find the answer for each word problem.**

1) Tickets to a movie cost $6 for adults and $4 for students. A group of friends purchased 9 tickets for $50.00. How many adults ticket did they buy? _____

2) At a store, Eva bought two shirts and five hats for $77.00. Nicole bought three same shirts and four same hats for $84.00. What is the price of each shirt? _____

3) A farmhouse shelters 10 animals, some are pigs, and some are ducks. Altogether there are 36 legs. How many pigs are there? _____

4) A class of 85 students went on a field trip. They took 24 vehicles, some cars and some buses. If each car holds 3 students and each bus hold 16 students, how many buses did they take? _____

5) A theater is selling tickets for a performance. Mr. Smith purchased 8 senior tickets and 10 child tickets for $248 for his friends and family. Mr. Jackson purchased 4 senior tickets and 6 child tickets for $132. What is the price of a senior ticket? $_____

6) The difference of two numbers is 15. Their sum is 33. What is the bigger number? $_____

7) The sum of the digits of a certain two-digit number is 7. Reversing its digits increase the number by 9. What is the number? _____

8) The difference of two numbers is 11. Their sum is 25. What are the numbers? _____

9) The length of a rectangle is 5 meters greater than 2 times the width. The perimeter of rectangle is 28 meters. What is the length of the rectangle? _____

10) Jim has 23 nickels and dimes totaling $2.40. How many nickels does he have? _____

Algebra 2 Workbook

Systems of 3 Variable Equations

✎ **Solve each system of equations.**

1) $x = 3y - 3z + 8$ $\quad x = \underline{}$
   $z = 4x + 5y - 14$ $\quad y = \underline{}$
   $3y + 2z = 14$ $\quad z = \underline{}$

2) $6x - 6y = -12$ $\quad x = \underline{}$
   $2z = -6x - 6y + 18$ $\quad y = \underline{}$
   $-8x + 10y + 2z = 16$ $\quad z = \underline{}$

3) $4x - 8z = 40$ $\quad x = \underline{}$
   $-6x + 2y - 8z = 40$ $\quad y = \underline{}$
   $-8x + 4y + 6z = -30$ $\quad z = \underline{}$

4) $2x - 4y + 2z = -12$ $\quad x = \underline{}$
   $2x + 10z = -24$ $\quad y = \underline{}$
   $-2x + 12y + 8z = 6$ $\quad z = \underline{}$

5) $x - y - 2z = -6$ $\quad x = \underline{}$
   $3x + 2y = -25$ $\quad y = \underline{}$
   $-4x + y - z = 12$ $\quad z = \underline{}$

6) $6x - y + 3z = -9$ $\quad x = \underline{}$
   $5x + 5y - 5z = 20$ $\quad y = \underline{}$
   $3x - y + 4z = -5$ $\quad z = \underline{}$

7) $-5x + 3y + 6z = 4$ $\quad x = \underline{}$
   $-3x + y + 5z = -5$ $\quad y = \underline{}$
   $-4x + 2y + z = 13$ $\quad z = \underline{}$

8) $-6x + 5y + 2z = -11$ $\quad x = \underline{}$
   $-2x + y + 4z = -9$ $\quad y = \underline{}$
   $4x - 5y + 5z = -4$ $\quad z = \underline{}$

9) $4x + 4y + z = 24$ $\quad x = \underline{}$
   $2x - 4y + z = 0$ $\quad y = \underline{}$
   $5x - 4y - 5z = 12$ $\quad z = \underline{}$

10) $-10x + 10y + 6z = -46$ $\quad x = \underline{}$
    $-10x + 6y - 6z = -22$ $\quad y = \underline{}$
    $-12x + 12z = -24$ $\quad z = \underline{}$

WWW.MathNotion.Com

Algebra 2 Workbook

Answers of Worksheets – Chapter 1

Relation and Functions

1) No, $D_f = \{2, 4, 8, 9, 10\}$, $R_f = \{1, 0, 5, 7, 14\}$
2) Yes, $D_f = \{2, 1, -1, 7, 9\}$, $R_f = \{3, 0, -4, 5\}$
3) Yes, $D_f = (-\infty, \infty)$, $R_f = \{2, -\infty)$
4) No, $D_f = \{6, 3, 2, 5\}$, $R_f = \{-8, -2, 4, 0, 9\}$
5) No, $D_f = [-2, \infty)$, $R_f = (-\infty, \infty)$
6) No, $D_f = \{3, 5, 6\}$, $R_f = \{0, 7, 1, 9\}$

Finding Slope

1) 2
2) −1
3) 4
4) −3
5) 7
6) −4
7) 7
8) −8
9) −9
10) 8
11) $\frac{1}{5}$
12) $-\frac{3}{7}$
13) $\frac{1}{2}$
14) −1
15) $\frac{3}{8}$
16) $\frac{3}{11}$
17) 2
18) 1
19) $-\frac{5}{2}$
20) 3
21) −1
22) 3
23) 2
24) −3
25) −1
26) 3
27) 3
28) −11

Graphing Lines Using Line Equation

1) $y = x - 5$ 2) $y = -3x + 4$ 3) $x - 2y = 0$

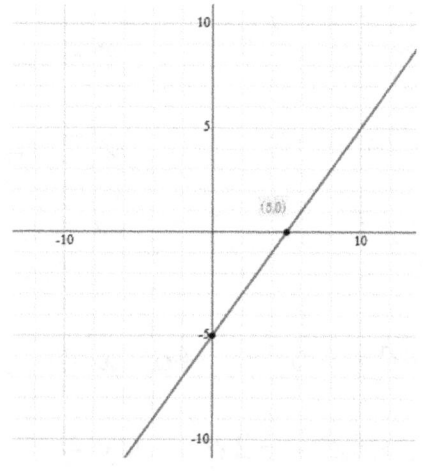
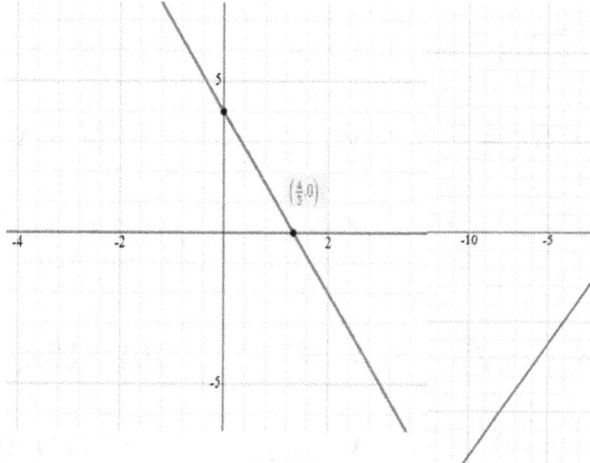

4) $x + y = -4$ 5) $4x + 3y = -2$ 6) $y - 3x + 6 = 0$

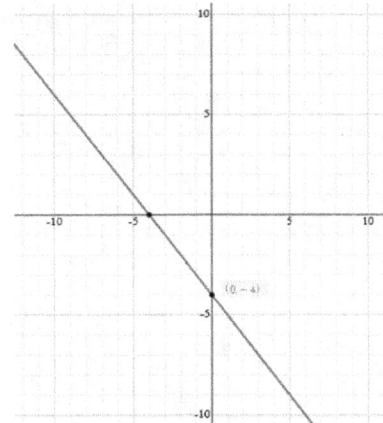 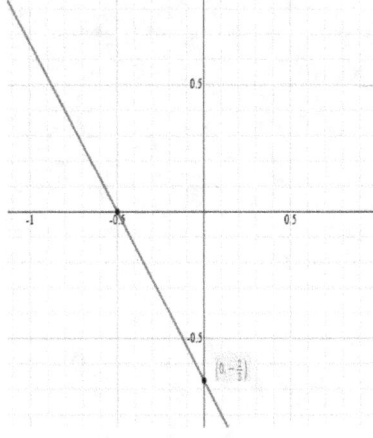 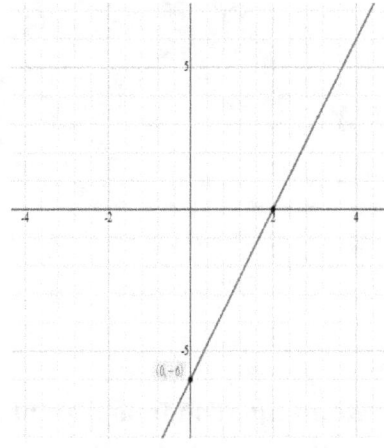

Writing Linear Equations

1) $y = 6x - 46$
2) $y = -3x + 34$
3) $y = x + 10$
4) $y = x - 4$
5) $y = -2x + 23$
6) $y = x - 15$
7) $y = -x + 20$
8) $y = -2x + 21$
9) $y = -3x + 42$
10) $y = -6x + 191$
11) $y = 8x + 57$
12) $y = 4x - 24$
13) $y = 4x - 34$
14) $y = -2x - 4$
15) $y = \frac{7}{6}x - \frac{5}{3}$
16) $y = -\frac{11}{3}x + \frac{28}{3}$
17) $y = 3x + 11$
18) $y = 5x + 15$
19) $y = 7x - 19$
20) $y = -3x - 1$
21) $y = -6x - 11$
22) $y = 5x + 22$

Graphing Linear Inequalities

1) $y > 3x - 5$ 2) $y < 2x + 1$ 3) $y \leq -4x - 5$

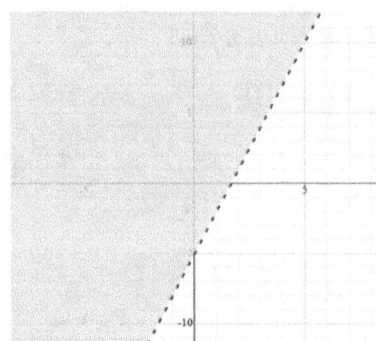 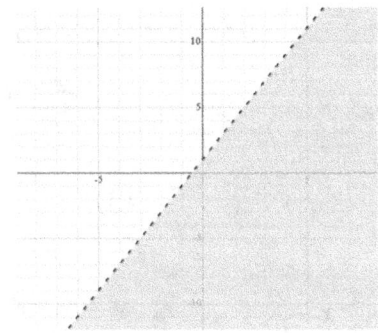 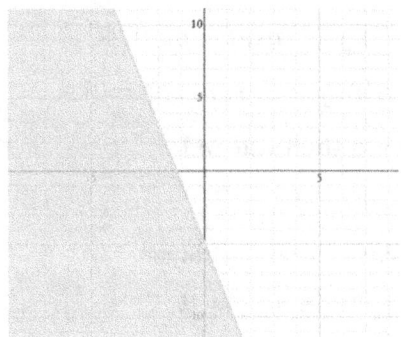

4) $2y \geq 12 + 4x$ 5) $-5y < x - 15$ 6) $3y \geq -9x + 6$

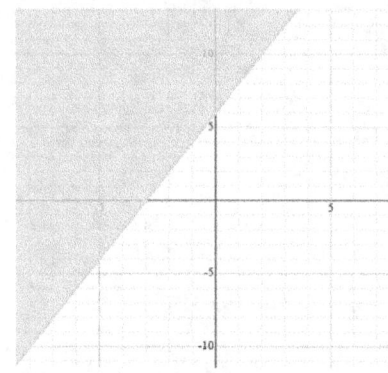
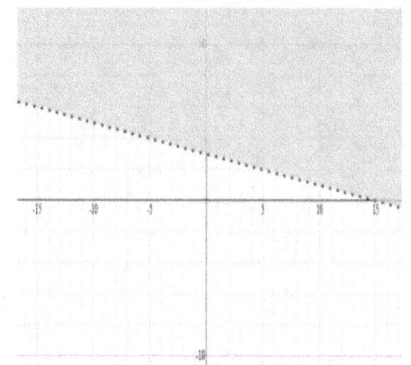
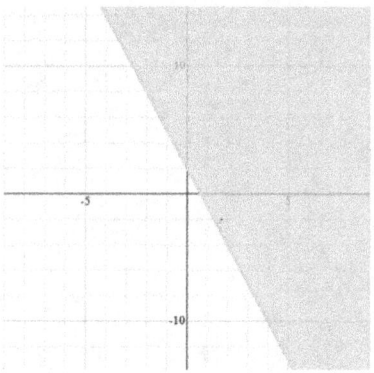

Write an equation from a graph.

1) $y = 4x + 2$ 2) $y = 2x + 6$ 3) $y = 8x$ 4) $y = 3x + 1$

Rate of change

1) 14   2) -10 3) 2

x–intercept and y–intercept

1) $y - intercept = 6$ $x - intercept = 3$

2) $y - intercept = 4$ $x - intercept = -4$

3) $y - intercept = -18$ $x - intercept = 6$

4) $y - intercept = -6$ $x - intercept = -6$

5) $y - intercept = -\frac{8}{5}$ $x - intercept = \frac{8}{7}$

6) $y - intercept = -\frac{12}{5}$ $x - intercept = 3$

7) $y - intercept = -\frac{15}{4}$ $x - intercept = -\frac{5}{4}$

8) $y - intercept =$ undefind $x - intercept = 4$

9) $y - intercept = 4$ $x - intercept =$ undefind

10) $y - intercept = 6$ $x - intercept = 10$

Find the value of b

11) -11 12) 1   13) 3   14) 2

Slope–intercept form

1) $y = 15x + 7$ 4) $y = -\frac{1}{2}x - \frac{3}{2}$ 7) $y = 3x + 6$ 10) $y = -\frac{5}{2}x + 4$

2) $y = -5x - 12$ 5) $y = \frac{2x}{5} - 3$ 8) $y = \frac{6}{5}x + \frac{36}{5}$ 11) $y = \frac{3}{2}x + 4$

3) $y = -\frac{1}{3}x + 2$ 6) $y = \frac{7}{5}x + 2$ 9) $y = 12x - 20$ 12) $y = -\frac{12}{7}x - 3$

Algebra 2 Workbook

13) $y = \frac{1}{3}x - \frac{4}{3}$ 14) $y = 2x + 11$ 15) $y = \frac{3}{4}x - \frac{7}{2}$ 16) $y = -\frac{3}{2}x - 2$

Point–slope form

1) $m = 3, (-5, 0)$
2) $m = \frac{1}{4}, (3, -2)$
3) $m = -\frac{5}{2}, (0, -1)$
4) $m = 3, (4, 4)$
5) $m = \frac{6}{10}, (-7, -5)$
6) $m = -2, (0, 6)$
7) $m = -2, (9, 10)$
8) $m = 0, (0, -18)$
9) $m = 6, (-1, -19)$
10) $m = -3, (-2, 14)$
11) $y + 7 = 5(x - 9)$
12) $y - 5 = \frac{1}{2}(x + 2)$
13) $y + 4 = -3x$
14) $y - b = n(x + a)$
15) $y - 2 = 4(x + 8)$
16) $y - 1 = -4(x - 6)$
17) $y - 12 = \frac{1}{6}(x + 7)$
18) $y - 13 = 0$
19) $y - 2 = \frac{1}{7}\left(x + \frac{1}{2}\right)$
20) $y = -5x$

Graphing Line of Equation

1) $y = 3x - 2$

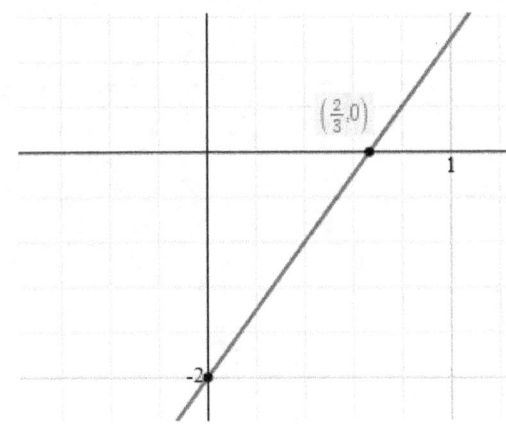

2) $y = -\frac{1}{2}x + \frac{3}{2}$

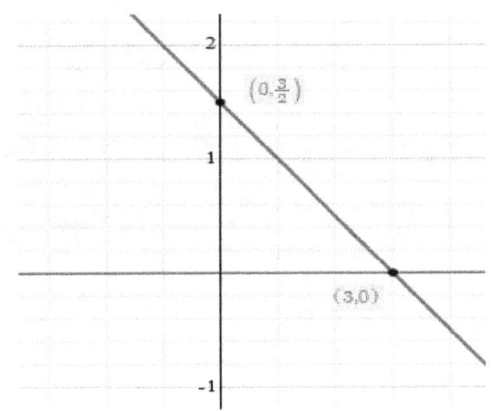

3) $4x - 5y = 12$

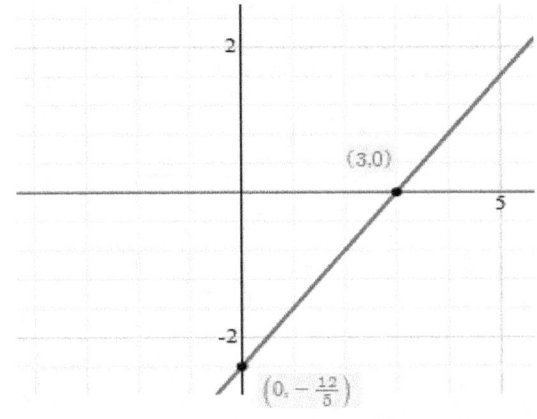

4) $-3x - y = 5$

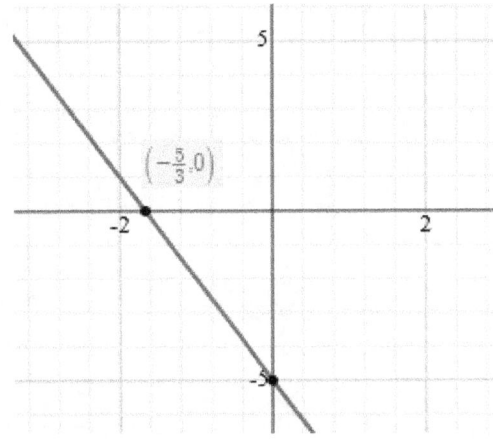

www.MathNotion.Com

Algebra 2 Workbook

Equation of parallel or perpendicular line.

1) $y = -\frac{1}{2}x - 2\frac{1}{2}$
2) $y = x + 5$
3) $y = \frac{1}{4}x + \frac{7}{4}$
4) $y = 2x + 1$
5) $y = 8$
6) $y = -2x + 6$
7) $y = \frac{2}{3}x + 2$
8) $y = 4x + 7$
9) $y = -\frac{1}{5}x - \frac{4}{5}$
10) $y = -\frac{1}{5}x - \frac{18}{5}$
11) $y = \frac{1}{2}x + 2$
12) $y = -\frac{2}{3}x + \frac{5}{12}$
13) $x = 4$
14) $y = -5$
15) $y = -4x - 25$
16) $y = \frac{1}{4}x + \frac{7}{12}$
17) $y = -\frac{4}{7}x - \frac{60}{7}$
18) $y = -x - 3$
19) $y = -\frac{4}{9}x - \frac{106}{9}$
20) $y = -5x + 27$

Equations of horizontal and vertical lines

1) $y = -2$

2) $y = 1$

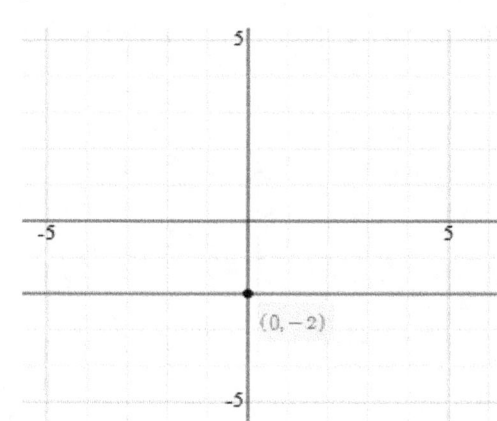
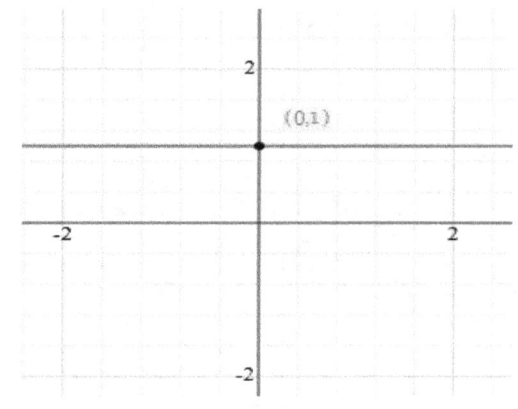

3) $x = -1$

4) $x = 2$

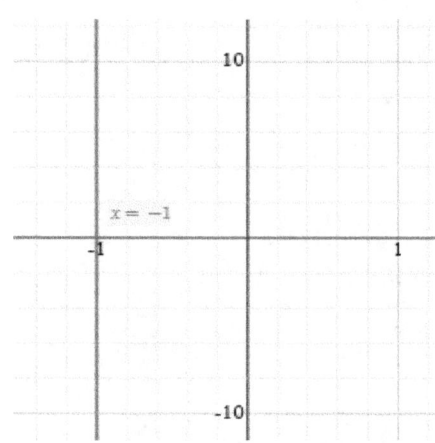
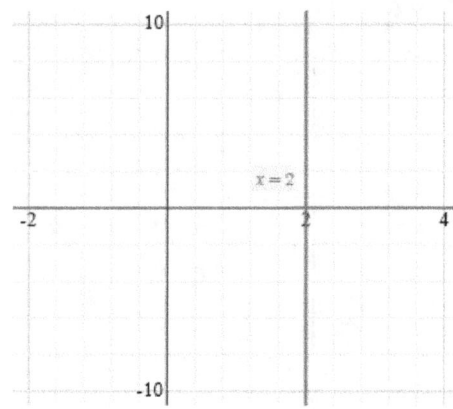

Algebra 2 Workbook

Graphing Absolute Value Equations

1) $y = |x + 4|$

2) $y = |x - 1|$

3) $y = -|x| - 1$

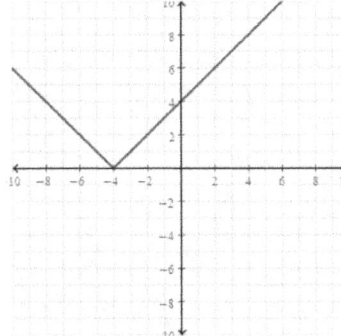
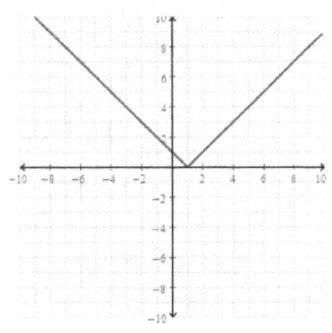
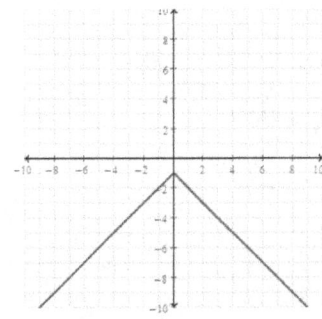

4) $y = |x - 2|$

5) $y = -|x - 2|$

6) $y = -2|2x + 2| + 4$

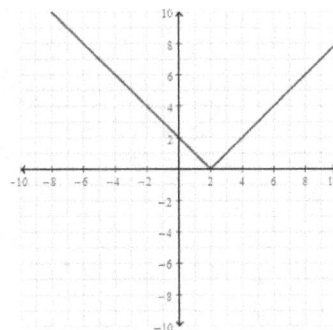
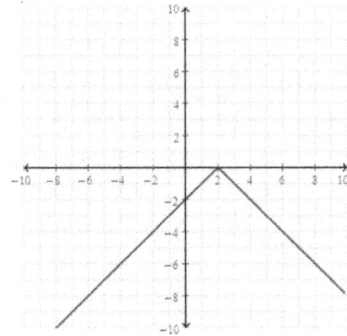
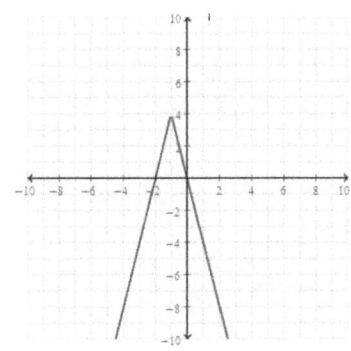

Systems of Equations

1) $x = 1, y = 2$

2) $x = 3, y = 2$

3) $x = -\frac{33}{2}$

4) $x = -\frac{1}{5}, y = \frac{26}{5}$

5) $x = -4, y = -3$

6) $x = 1, y = -2$

7) $x = 1, y = 2$

8) $x = \frac{9}{5}, y = \frac{17}{5}$

9) $x = -110, y = -85$

10) $x = -\frac{15}{4}, y = \frac{29}{4}$

11) $x = \frac{5}{3}, y = \frac{2}{3}$

12) $x = -\frac{15}{4}, y = \frac{29}{4}$

13) $x = \frac{33}{19}, y = -\frac{15}{19}$

14) $x = -2, y = -4$

15) $x = -\frac{2}{7}, y = \frac{27}{14}$

16) $x = \frac{91}{17}, y = -\frac{132}{17}$

Systems of Equations Word Problems

1) 7

2) $16

3) 8

4) 1

5) $21

6) 24

7) 43

8) 18, 7

9) 11 meters

10) 18

Systems of 3 variable equations

1) $(2, 2, 4)$

2) $(1, 3, -3)$

3) $(0, 0, -5)$

4) $(3, 3, -3)$

5) $(-5, -5, 3)$

6) $(-1, 6, 1)$

7) $(-2, 4, -3)$

8) $(4, 3, -1)$

9) $(4, 2, 0)$

10) $(1, -3, -1)$

Chapter 2: Monomials and polynomials

Topics that you will practice in this chapter:

- ✓ GCF of Monomials
- ✓ Factoring Quadratics
- ✓ Factoring by Grouping
- ✓ GCF and Powers of Monomials
- ✓ Writing Polynomials in Standard Form
- ✓ Simplifying Polynomials
- ✓ Adding and Subtracting Polynomials
- ✓ Multiplying a Polynomial and a Monomial
- ✓ Multiplying Binomials
- ✓ Factoring Trinomials
- ✓ Operations with Polynomials

Mathematics is, as it were, a sensuous logic, and relates to philosophy as do the arts, music, and plastic art to poetry. — *K. Shegel*

GCF of Monomials

✏️ **Find the GCF of each set of monomials.**

1) $39x, 30xy$

2) $60a, 56a^2$

3) $18x^2, 54x^2$

4) $36x^2, 21x^3$

5) $20a^2, 30a^2b$

6) $80a^3, 30a^2b$

7) $54x^3, 36x^3$

8) $33x, 44y^2x$

9) $15x^2, 12, 48$

10) $10v^3, 45v^3, 35v$

11) p^2q^2, pqr

12) $15m^2n, 25m^2n^2$

13) $12x^2yz, 3xy^2$

14) $22m^5n^2, 11m^2n^4$

15) $16x^3y, 8x^2$

16) $14ab^5, 7a^2b^2c$

17) $12t^7u^2, 18t^3u^7$

18) $18t, 48t^4$

19) $18r^3t, 26qr^2t^4$

20) $11a^4b^3, 44a^2b^5$

21) $16f, 21ab^2$

22) $12a^2b^2c^2, 20abc$

23) $18ab, 9ab$

24) $22m^5n^2, 11m^2n^4$

25) $4xy, 2x^2$

26) $x^3yz^2, 2x^3yz^3$

27) $140x, 140y^2, 80y^2$

28) $24a, 36a, 24ab^2$

29) $10x^3, 45x^3, 35x$

30) $105a, 30ab, 75a$

Factoring Quadratics

✏️ **Factor each completely.**

1) $x^2 - 16x + 63 =$

2) $m^2 - 9m + 8 =$

3) $p^2 - 5p - 14 =$

4) $2b^2 + 17b + 21 =$

5) $a^2 + 5a + 4 =$

6) $a^2 + 2a - 15 =$

7) $4n^2 + 12n + 9 =$

8) $t^2 + 2t - 19 =$

9) $3x^3 + 21x^2 + 36x =$

10) $x^2 + 5x + 6 =$

11) $9r^2 - 5r - 10 =$

12) $30n^2 b - 87nb + 30b =$

13) $7x^2 - 32x - 60 =$

14) $3b^3 - 5b^2 + 2b =$

15) $10m^2 + 89m - 9 =$

16) $4x^3 + 43x^2 + 30x =$

17) $9x^2 + 7 - 56 =$

18) $p^2 - 5p - 14 =$

19) $x^2 - 7x - 18 =$

20) $7x^2 - 31x - 20 =$

21) $6n^2 + 7n - 49 =$

22) $-6x^2 - 25x - 25 =$

23) $6x^2 + 5x - 6 =$

24) $16x^2 + 60x - 100 =$

25) $4x^2 - 35x + 49 =$

26) $5x^2 - 18x + 9 =$

27) $9n^2 + 66n + 21 =$

28) $3x^2 - 8x + 4 =$

29) $6x^2 - 36xy =$

30) $-6x^3 - 23x^2 y - 10y^2 x =$

31) $9a^2 + 9ab - 4b^2 =$

32) $4x^2 + 4xy - 35y^2 =$

33) $7x^2 y - 27xy^2 + 18y^3 =$

34) $-2x^2 + 8xy + 64y^2 =$

35) $25mp^2 - 45mp =$

36) $14b^2 + 142b + 144 =$

37) $5x^2 + 85xy + 350y^2 =$

38) $7x^2 + 9xy =$

Factoring by Grouping

✎ Factor each completely.

1) $28xy - 7k - 49x + 4ky =$

2) $7xy - 3n - x + 21ny =$

3) $56n^3 + 64n^2 + 70n + 80 =$

4) $32u^2v - 12u^3m + 48u^4 - 8umv =$

5) $70n^4 + 40n^3 + 28n^2 + 16n =$

6) $45uv - 125bu - 75u^2 + 75bv =$

7) $x^3 + 7x^2 + 6x + 42 =$

8) $6x^3 + 36x^2 + 30x + 180 =$

9) $6m^3 - 30m^2 + 30m - 150 =$

10) $2x^3 - 4x^2 - 10x + 20 =$

11) $24p^3 + 15p^2 - 56p - 35 =$

12) $42mc + 36md - 7n^2c - 6n^2d =$

13) $28x^4 + 112x^2 - 21x^2 - 84x =$

14) $15xw + 18xk + 25yw + 30k =$

15) $56xy - 35x + 16ry - 10r =$

16) $4xy + 6 - x - 24y =$

17) $192x3 + 72x2 + 144x + 54 =$

18) $8x3 - 8x2 + 14x - 14 =$

19) $20x^3 + 5x^2 + 28x + 7 =$

20) $100x^3 + 160x^2 - 60x - 96 =$

GCF and Powers of Monomials

✏ Find the GCF of each pairs of expressions.

1) $54x^3, 36x^3$

2) 2) $33x, 44y^2x$

3) $15x^2, 12, 48$

4) 4) $10v^3, 45v^3, 35v$

5) p^2q^2, pqr

6) 6) $15m^2n, 25m^2n^2$

7) $12x^2yz, 3xy^2$

8) 8) $22m^5n^2, 11m^2n^4$

9) $16x^3y, 8x^2$

10) $14ab^5, 7a^2b^2c$

11) $12t^7u^2, 18t^3u^7$

12) 12) $18t, 48t^4$

13) $18r^3t, 26qr^2t^4$

14) 14) $11a^4b^3, 44a^2b^5$

15) $16f, 21ab^2$

16) 16) $12a^2b^2c^2, 20abc$

17) $18ab, 9ab$

18) 18) $22m^5n^2, 11m^2n^4$

19) $4xy, 2x^2$

20) 20) $x^3yz^2, 2x^3yz^3$

✏ Simplify.

21) $(3x^4)^7$

22) $(4y^22y^3y)^2$

23) $(3x^2\,2x^2)^3$

24) $(8x^4y^3)^6$

25) $(3y^25y^2)^2$

26) $(6x^3y)^3$

27) $(8x^2x^23n)^2$

28) $(7xy^6)^3$

29) $(9x^3y^2)^4$

30) $(10y^3y^2)^3$

31) $(6x^2x^6)^3$

32) $(3x^74x^3k^2)^2$

33) $(4y^54y^2)^2$

34) $(5x2x^3)^3$

35) $(4y^3)^3$

36) $(y^3y^3y^2)^3$

37) $(4y^2y)^3$

38) $(6xy^6)^3$

Writing Polynomials in Standard Form

✎ **Write each polynomial in standard form.**

1) $9x - 7x =$

2) $-6 + 15x - 15x =$

3) $3x^2 - 11x^3 =$

4) $18 + 19x^3 - 14 =$

5) $3x^2 + 9x - 4x^5 =$

6) $-7x^3 + 12x^7 =$

7) $9x + 6x^2 - 2x^6 =$

8) $-5x^3 + x - 9x^4 =$

9) $8x^2 + 34 - 21x =$

10) $8 - 7x + 11x^4 =$

11) $25x^3 + 45x - 13x^4 =$

12) $17 + 9x^2 - 2x^3 =$

13) $18x^2 - 8x + 8x^3 =$

14) $9x^4 - 4x^2 - 10x^5 =$

15) $-41 + 7x^2 - 8x^4 =$

16) $8x^2 - 7x^5 + 3x^3 - 12 =$

17) $4x^2 - 9x^5 + 12 - 8x^4 =$

18) $-2x^5 + 6x - 9x^2 - 7x =$

19) $14x^5 + 7x^4 - 8x^5 - 8x^2 =$

20) $2x^3 - 15x^4 + 9x^3 + 3x^8 =$

21) $7x^4 - 16x^5 - 9x^2 + 10x^4 =$

22) $5x^2 + 6x^5 + 37x^3 - 9x^5 =$

23) $3x(2x + 5 - 6x^2) =$

24) $12x(x^6 + 2x^3) =$

25) $6x(x^2 + 8x + 4) =$

26) $8x(3 - 2x + 4x^3) =$

27) $7x(2x^3 - 2x^2 + 2) =$

28) $5x(5x^5 + 4x^4 - 1) =$

29) $x(4x^3 + 52x^4 + 2x) =$

30) $6x(3x - 4x^4 + 7x^2) =$

Algebra 2 Workbook

Simplifying Polynomials

✎ **Simplify each expression.**

1) $3(x - 12) =$

2) $5x(2x - 4) =$

3) $7x(5x - 1) =$

4) $6x(3x + 2) =$

5) $5x(2x - 7) =$

6) $9x(x + 8) =$

7) $(3x - 8)(x - 3) =$

8) $(x - 9)(3x + 4) =$

9) $(x - 8)(x - 5) =$

10) $(3x + 4)(3x - 4) =$

11) $(5x - 8)(5x - 2) =$

12) $7x^2 + 7x^2 - 6x^4 =$

13) $5x - 2x^2 + 7x^3 + 10 =$

14) $8x + 2x^2 - 5x^3 =$

15) $15x + 4x^5 - 8x^2 =$

16) $-4x^2 + 7x^5 + 11x^4 =$

17) $-14x^2 + 8x^3 - 2x^4 + 5x =$

18) $14 - 5x^2 + 6x^2 - 10x^3 + 17 =$

19) $x^2 - 9x + 2x^3 + 15x - 10x =$

20) $14 - 8x^2 + 4x^2 - 9x^3 + 1 =$

21) $-4x^5 + 2x^4 - 18x^2 + 2x^5 =$

22) $(3x^3 - 5) + (3x^3 - 2x^3) =$

23) $4(3x^5 - 3x^3 - 6x^5) =$

24) $-4(x^5 + 8) - 4(12 - x^5) =$

25) $7x^2 - 9x^3 - 2x + 14 - 5x^2 =$

26) $10 - 5x^2 + 3x^2 - 4x^3 + 4 =$

27) $(8x^2 - 2x) - (5x - 5 - 4x^2) =$

28) $4x^4 - 8x^3 - x(3x^2 + 5x) =$

29) $4x + 8x^2 - 10 - 2(x^2 - 1) =$

30) $5 - 3x^2 + (6x^4 - 2x^2 + 8x^4) =$

31) $-(x^5 + 8) - 7(4 + x^5) =$

32) $(4x^3 - x) - (x - 6x^3) =$

WWW.MathNotion.Com

Algebra 2 Workbook

Adding and Subtracting Polynomials

✎ **Add or subtract expressions.**

1) $(-x^3 - 3) + (4x^3 + 2) =$

2) $(3x^2 + 4) - (6 - x^2) =$

3) $(x^3 + 4x^2) - (5x^3 + 15) =$

4) $(3x^3 - 2x^2) + (2x^2 - x) =$

5) $(10x^3 + 14x) - (14x^3 + 7) =$

6) $(5x^2 - 7) + (3x^2 + 7) =$

7) $(9x^3 + 4) - (10 - 5x^3) =$

8) $(x^2 + 2x^3) - (2x^3 + 5) =$

9) $(8x^2 - x) + (5x - 4x^2) =$

10) $(17x + 10) - (2x + 10) =$

11) $(12x^4 - 4x) - (x - 3x^4) =$

12) $(3x - x^4) - (7x^4 + 8x) =$

13) $(7x^3 - 6x^5) - (4x^5 - 2x) =$

14) $(x^3 - 7) + (4x^3 + 8x^5) =$

15) $(6x^2 + 5x^4) - (x^4 - 9x^2) =$

16) $(-4x^2 - 4x) + (7x - 8x^2) =$

17) $(x - 6x^4) - (15x^4 + 2x) =$

18) $(4x - 3x^4) - (2x^4 - 3x^3) =$

19) $(7x^3 - 7) + (6x^3 - 6x^2) =$

20) $(9x^5 + 7x^4) - (x^4 - 5x^5) =$

21) $(-4x^2 + 11x^4 + 2x^3) + (20x^3 + 4x^4 + 12x^2) =$

22) $(5x^2 - 5x^4 - 5x) - (-4x^2 - 5x^4 + 5x) =$

23) $(12x + 36x^3 - 10x^4) + (20x^3 + 10x^4 - 7x) =$

24) $(2x^5 - 4x^3 - 5x) - (2x^2 + 7x^3 - 2x) =$

25) $(14x^3 - 4x^5 - x) - (-4x^3 - 12x^5 + 9x) =$

26) $(-5x^2 + 12x^4 + x^3) + (10x^3 + 17x^4 + 7x^2) =$

WWW.MathNotion.Com

Multiplying a Polynomial and a Monomial

✏ **Find each product.**

1) $2x(x + 4) =$

2) $3(8 - x) =$

3) $5x(3x + 4) =$

4) $x(-2x + 5) =$

5) $7x(3x - 3) =$

6) $3(2x - 5y) =$

7) $6x(7x - 3) =$

8) $x(12x + 5y) =$

9) $5x(x + 6y) =$

10) $11x(4x + 5y) =$

11) $8x(4x + 2) =$

12) $12x(x - 15y) =$

13) $9x(5x - 3y) =$

14) $8x(5x - 2y + 5) =$

15) $9x(2x^2 + 7y^2) =$

16) $8x(9x + 6y) =$

17) $2(3x^5 - 2y^5) =$

18) $4x(-x^2y + 2y) =$

19) $-3(2x^3 - 3xy + 9) =$

20) $2(x^2 - 2xy - 4) =$

21) $7x(4x^3 - xy + 2x) =$

22) $-9x(-2x^3 - 2x + 7xy) =$

23) $6(x^2 + 3xy - 8y^2) =$

24) $5x(7x^3 - x + 8) =$

25) $7(x^{24} - 4x - 6) =$

26) $x^2(-3x^3 + 4x + 7) =$

27) $x^2(2x^3 + 10 - 5x) =$

28) $4x^4(3x^3 - 2x + 8) =$

29) $5x^2(x^4 - 5xy + 2y^3) =$

30) $4x^2(7x^4 - 2x + 11) =$

31) $7x^3(3x^3 + 5x - 7) =$

32) $4x(x^2 - 8xy + 7y^3) =$

Multiplying Binomials

✎ **Find each product.**

1) $(x + 5)(x + 1) =$

2) $(x - 3)(x + 7) =$

3) $(x - 1)(x - 9) =$

4) $(x + 3)(x + 8) =$

5) $(x - 4)(x - 11) =$

6) $(x + 5)(x + 6) =$

7) $(x - 8)(x + 7) =$

8) $(x - 3)(x - 2) =$

9) $(x + 8)(x + 11) =$

10) $(x - 3)(x + 5) =$

11) $(x + 8)(x + 8) =$

12) $(x + 2)(x + 7) =$

13) $(x - 9)(x + 4) =$

14) $(x - 10)(x + 10) =$

15) $(x + 24)(x + 2) =$

16) $(x + 9)(x + 13) =$

17) $(x - 7)(x + 7) =$

18) $(x - 5)(x + 2) =$

19) $(3x + 4)(x + 5) =$

20) $(x - 8)(5x + 2) =$

21) $(x - 9)(4x + 9) =$

22) $(2x - 7)(3x - 2) =$

23) $(x - 4)(x + 11) =$

24) $(5x - 6)(2x + 4) =$

25) $(4x - 9)(x + 7) =$

26) $(8x - 5)(2x + 2) =$

27) $(3x + 9)(7x + 4) =$

28) $(6x - 8)(4x + 4) =$

29) $(4x + 5)(5x - 8) =$

30) $(8x - 1)(8x + 4) =$

31) $(9x + 4)(3x - 6) =$

32) $(4x^2 + 12)(4x^2 - 12) =$

Factoring Trinomials

✎ **Factor each trinomial.**

1) $x^2 + 12x + 35 =$

2) $x^2 - 8x + 12 =$

3) $x^2 + 11x + 10 =$

4) $x^2 - 12x + 27 =$

5) $x^2 - 16x + 15 =$

6) $x^2 - 13x + 40 =$

7) $x^2 + 15x + 44 =$

8) $x^2 + x - 72 =$

9) $x^2 - 81 =$

10) $x^2 - 17x + 70 =$

11) $x^2 + 8x - 48 =$

12) $x^2 + 5x - 104 =$

13) $x^2 - 7x - 18 =$

14) $x^2 + 22x + 121 =$

15) $3x^2 - 3x - 36 =$

16) $2x^2 - 35x + 75 =$

17) $14x^2 + 11x - 15 =$

18) $8x^2 - 12x - 20 =$

19) $15x^2 + 16x + 4 =$

20) $24x^2 + 2x - 1 =$

✎ **Calculate each problem.**

21) The area of a rectangle is $x^2 - 3x - 40$. If the width of rectangle is $x - 8$, what is its length? _____

22) The area of a parallelogram is $12x^2 + 7x - 10$ and its height is $4x + 5$. What is the base of the parallelogram? _____

23) The area of a rectangle is $10x^2 - 43x + 28$. If the width of the rectangle is $5x - 4$, what is its length? _____

Algebra 2 Workbook

Operations with Polynomials

✍ **Find each product.**

1) $2(4x + 1) =$ _____

2) $5(2x + 7) =$ _____

3) $4(6x - 5) =$ _____

4) $-4(7x - 8) =$ _____

5) $3x^2(8x + 4) =$ _____

6) $6x^2(2x - 9) =$ _____

7) $5x^3(-x + 4) =$ _____

8) $-5x^4(4x - 9) =$ _____

9) $6(x^2 + 7x - 3) =$ _____

10) $4(3x^2 - 2x + 6) =$ _____

11) $9(3x^2 + 8x + 2) =$ _____

12) $7x(x^2 + 5x + 3) =$ _____

13) $(7x + 2)(2x - 5) =$ _____

14) $(8x + 5)(3x - 8) =$ _____

15) $(4x + 2)(6x - 1) =$ _____

16) $(5x - 4)(5x + 9) =$ _____

✍ **Calculate each problem.**

17) The measures of two sides of a triangle are $(2x + 8y)$ and $(5x - 3y)$. If the perimeter of the triangle is $(11x + 6y)$, what is the measure of the third side? _____

18) The height of a triangle is $(8x + 2)$ and its base is $(2x - 6)$. What is the area of the triangle? _____

19) One side of a square is $(4x + 3)$. What is the area of the square? _____

20) The length of a rectangle is $(7x - 9y)$ and its width is $(13x + 9y)$. What is the perimeter of the rectangle? _____

21) The side of a cube measures $(x + 2)$. What is the volume of the cube? _____

22) If the perimeter of a rectangle is $(24x + 10y)$ and its width is $(4x + 3y)$, what is the length of the rectangle? _____

WWW.MathNotion.Com

Answers of Worksheets – Chapter 6

GCF of Monomials

1) $3x$	11) pq	21) no
2) 4	12) $5m^2n$	22) $4abc$
3) $18x^2$	13) $3xy$	23) $9ab$
4) $3x^2$	14) $11m^2n$	24) $11m^2n^2$
5) $10a^2$	15) $8x^2$	25) $2x$
6) $10a^2$	16) $7ab^2$	26) x^3yz^2
7) $18x^3$	17) $6t^3u^2$	27) 20
8) $11x$	18) $8t$	28) $12a$
9) 3	19) $2r^2t$	29) $5x$
10) $5v$	20) $11a^2b^3$	30) $15a$

Factoring Quadratics

1) $(x-9)(x-7)$	18) $p^2 - 5p - 14$
2) $(m-1)(m-8)$	19) $x^2 - 7x - 18$
3) $(p+2)(p-7)$	20) $7x^2 - 31x - 20$
4) $(2b+3)(b+7)$	21) $(3n-7)(2x+7)$
5) $a^2 + 5a + 4$	22) $-(2x+5)(3x+5)$
6) $a^2 + 2a - 15$	23) $(2x+3)(3x-2)$
7) $4n^2 + 12n + 9$	24) $4(x+5)(4x-5)$
8) $t^2 + 2t - 19$	25) $(x-7)(4x-7)$
9) $3x^3 + 21x^2 + 36x$	26) $(5x-3)(x-3)$
10) $x^2 + 5x + 6$	27) $3(3n+1)(n+7)$
11) $9r^2 - 5r - 10$	28) $(3x-2)(x-2)$
12) $30n^2b\ 87nb + 30b$	29) $6x(x-6y)$
13) $7x^2 - 32x - 60$	30) $-x(2x+y)(3x+10y)$
14) $3b^3 - 5b^2 + 2b$	31) $(3a+4b)(3a-b)$
15) $10m^2 + 89m - 9$	32) $(2x+7y)(2x-5y)$
16) $4x^3 + 43x^2 + 30x$	33) $y(7x-6y)(x-3y)$
17) $9x^2 + 7x - 56$	34) $-2(x-8y)(x+4y)$

Algebra 2 Workbook

35) $5mp(5p - 9)$

36) $2(7b + 8)(b + 9)$

37) $5(x + 10y)(x + 7y)$

38) $x(7x + 9y)$

Factoring by Grouping

1) $(7x + k)(4y - 7)$

2) $(x + 3n)(7y - 1)$

3) $2(4n^2 + 5)(7n + 8)$

4) $4u(4u - m)(2v + 3u^2)$

5) $2n(5n^2 + 2)(7n + 4)$

6) $5(3u + 5b)(3v - 5u)$

7) $(x^2 + 6)(x + 7)$

8) $6(x^2 + 5)(x + 6)$

9) $6(m^2 + 5)(m - 5)$

10) $2(x^2 - 5)(x - 2)$

11) $(3p^2 - 7)(8p + 5)$

12) $(6m - n^2)(7c + 6d)$

13) $7x(4x^2 - 3)(x + 4)$

14) $(3x + 5y)(5w + 6k)$

15) $(7x + 2r)(8y - 5)$

16) $(x - 6)(4y - 1)$

17) $(4x^2 + 1)(3x - 5)$

18) $2(4x^2 + 7)(x - 1)$

19) $(5x^2 + 7)(4x + 1)$

20) $4(5x^2 - 3)(5x + 8)$

GCF and Powers of monomials

1) $10x^2$

2) 4

3) 10

4) $7xy$

5) $6y7x$

6) $3x$

7) 3

8) $7xy^2$

9) $18x^2$

10) $(15)x$

11) 27

12) 3

13) $20x$

14) $2x^2y$

15) xy

16) $2x^2y^2$

17) $6xy^4$

18) $5x^3y$

19) $2187x^{28}$

20) $72x^{12}$

21) $216x^{12}$

22) $262144x^{24}y^{18}$

23) $225y^{10}$

24) $216x^9y^3$

25) $576x^8n^2$

26) $343x^3y^{18}$

27) $6561x^{12}y^8$

28) $1000y^{15}$

29) $396x^{24}$

30) $144x^{100}k^4$

31) $256y^{14}$

32) $1000x^{12}$

33) $64y^9$

34) $27y^{18}$

35) $64y^9$

36) $216x^3y^{18}$

Writing Polynomials in Standard Form

1) $2x$

2) -6

3) $-11x^3 + 3x^2$

4) $19x^4 + 4$

5) $-4x^5 + 3x^2 + 9x$

6) $12x^7 - 7x^3$

7) $-2x^6 + 6x^2 + 9x$

8) $-9x^4 - 5x^3 + x$

Algebra 2 Workbook

9) $8x^2 - 21x + 34$
10) $11x^4 - 7x + 8$
11) $-13x^4 + 25x^3 + 45x$
12) $-2x^3 + 9x^2 + 17$
13) $8x^3 + 18x^2 - 8x$
14) $-10x^5 - 9x^4 - 4x^2$
15) $-8x^4 + 7x^2 - 41$
16) $-7x^5 + 3x^3 + 8x^2 - 12$
17) $-9x^5 - 8x^4 + 4x^2 + 12$
18) $-2x^5 - 9x^2 - x$
19) $6x^5 + 7x^4 - 8x^2$
20) $3x^8 - 15x^4 + 11x^2$
21) $-16x^5 + 17x^4 - 9x^2$
22) $-3x^5 + 37x^3 + 5x^2$
23) $-18x^3 + 6x^2 + 15x$
24) $12x^7 + 24x^4$
25) $6x^3 + 48x^2 + 24x$
26) $32x^4 - 16x^2 + 24x$
27) $14x^4 - 14x^3 + 14x$
28) $25x^6 + 20x^5 - 5x$
29) $52x^5 + 4x^4 + 2x^2$
30) $-24x^5 + 42x^3 + 18x^2$

Simplifying Polynomials

1) $3x - 36$
2) $10x^2 - 20x$
3) $35x^2 - 7x$
4) $18x^2 + 12x$
5) $10x^2 - 35x$
6) $9x^2 + 72x$
7) $3x^2 - 17x + 24$
8) $3x^2 - 23x - 36$
9) $x^2 - 13x + 40$
10) $9x^2 - 16$
11) $25x^2 - 50x + 16$
12) $-6x^4 + 14x^2$
13) $7x^3 - 2x^2 + 5x + 10$
14) $-5x^3 + 2x^2 + 8x$
15) $4x^5 - 8x^2 + 15x$
16) $7x^5 + 11x^4 - 4x^2$
17) $-2x^4 + 8x^3 - 14x^2 + 5x$
18) $-10x^3 + x^2 + 31$
19) $2x^3 + x^2 - 4x$
20) $-9x^3 - 4x^2 + 15$
21) $-2x^5 + 2x^4 - 18x^2$
22) $4x^3 - 5$
23) $-12x^5 - 12x^3$
24) -80
25) $-9x^3 + 2x^2 - 2x + 14$
26) $-4x^3 - 2x^2 + 14$
27) $12x^2 - 7x + 5$
28) $4x^4 - 11x^3 - 5x^2$
29) $6x^2 + 4x - 8$
30) $14x^4 - 5x^2 + 5$
31) $-8x^5 - 36$
32) $10x^3 - 2x$

Adding and Subtracting Polynomials

1) $3x^2 - 1$
2) $4x^2 - 2$
3) $-4x^3 + 4x^2 - 15$
4) $3x^3 - x$
5) $-4x^3 + 14x - 7$
6) $8x^2$

WWW.MathNotion.Com

Algebra 2 Workbook

7) $14x^3 - 6$
8) $x^2 - 5$
9) $4x^2 + 4x$
10) $15x$
11) $15x^4 - 5x$
12) $-8x^4 - 5x$
13) $-10x^5 + 7x^3 + 2x$

14) $5x^5 + 5x^3 - 7$
15) $4x^4 + 15x^2$
16) $-12x^2 + 3x$
17) $-21x^4 - x$
18) $-5x^4 + 3x^3 + 4x$
19) $13x^3 - 6x^2 - 7$
20) $14x^5 + 6x^4$

21) $15x^4 + 22x^3 + 8x^2$
22) $9x^2 - 10x$
23) $56x^3 + 5x$
24) $2x^5 - 11x^3 - 2x^2 - 3x$
25) $8x^5 + 18x^3 - 10x$
26) $29x^4 + 11x^3 + 2x^2$

Multiplying a Polynomial and a Monomial

1) $2x^2 + 8x$
2) $-3x + 24$
3) $15x^2 + 20x$
4) $-2x^2 + 5x$
5) $21x^2 - 21x$
6) $6x - 15y$
7) $42x^2 - 18x$
8) $12x^2 + 5xy$
9) $5x^2 + 30xy$
10) $44x^2 + 55xy$
11) $32x^2 + 16x$
12) $12 - 180xy$
13) $45x^2 - 27xy$
14) $40x^2 - 16xy + 40x$
15) $18x^3 + 63xy^2$
16) $72x^2 + 48xy$

17) $6x^5 - 2y^5$
18) $-4x^3y + 8xy$
19) $-6x^3 + 9xy - 27$
20) $2x^2 - 4xy - 8$
21) $28x^4 - 7x^2y + 14x^2$
22) $18x^4 + 18x^2 - 63x^2y$
23) $6x^2 + 18xy - 48y^2$
24) $35x^4 - 5x^2 + 40x$
25) $7x^{24} - 28x - 42$
26) $-3x^5 + 4x^3 + 7x^2$
27) $2x^5 - 5x^3 + 10x^2$
28) $12x^7 - 8x^5 + 32x^4$
29) $5x^6 - 25x^3y + 10x^2y^3$
30) $28x^6 - 8x^3 + 44x^2$
31) $21x^6 + 35x^4 - 49x^3$
32) $4x^3 - 32x^2y + 28xy^3$

Multiplying Binomials

1) $x^2 + 6x + 5$
2) $x^2 + 4x - 21$
3) $x^2 - 10x + 9$
4) $x^2 + 11x + 24$
5) $x^2 - 15x + 44$
6) $x^2 + 11x + 30$

7) $x^2 - x - 56$
8) $x^2 - 5x + 6$
9) $x^2 + 19x + 88$
10) $x^2 + 2x + 15$
11) $x^2 + 16x + 64$
12) $x^2 + 9x + 14$

Algebra 2 Workbook

13) $x^2 - 5x - 36$
14) $x^2 - 100$
15) $x^2 + 26x + 48$
16) $x^2 + 22x + 117$
17) $x^2 - 49$
18) $x^2 - 3x - 10$
19) $3x^2 + 19x + 20$
20) $5x^2 - 38x - 16$
21) $4x^2 - 27x - 81$
22) $6x^2 - 25x + 14$

23) $x^2 + 7x - 44$
24) $10x^2 + 8x - 24$
25) $4x^2 + 19x - 63$
26) $16x^2 + 6x - 10$
27) $21x^2 + 75x + 36$
28) $24x^2 - 8x - 32$
29) $20x^2 - 7x - 40$
30) $64x^2 + 24x - 4$
31) $27x^2 - 42x - 24$
32) $16x^4 - 144$

Factoring Trinomials

1) $(x+5)(x+7)$
2) $(x-2)(x-6)$
3) $(x+1)(x+10)$
4) $(x-9)(x-3)$
5) $(x-1)(x-15)$
6) $(x-5)(x-8)$
7) $(x+4)(x+11)$
8) $(x+9)(x-8)$

9) $(x-9)(x+9)$
10) $(x-7)(x-10)$
11) $(x-4)(x+12)$
12) $(x-8)(x+13)$
13) $(x+2)(x-9)$
14) $(x+11)(x+11)$
15) $(3x+9)(x-4)$
16) $(x-15)(2x-5)$

17) $(7x-5)(2x+3)$
18) $(2x-5)(4x+4)$
19) $(3x+2)(5x+2)$
20) $(6x-1)(4x+1)$
21) $(x+5)$
22) $(3x-2)$
23) $(2x-7)$

Operations with Polynomials

1) $8x + 2$
2) $10x + 35$
3) $24x - 20$
4) $-28x + 32$
5) $24x^3 + 12x^2$
6) $12x^3 - 54x^2$
7) $-5x^4 + 20x^3$
8) $-20x^5 + 45x^4$

9) $6x^2 + 42x - 18$
10) $12x^2 - 8x + 24$
11) $27x^2 + 72x + 18$
12) $7x^3 + 35x^2 + 21x$
13) $14x^2 - 31x - 10$
14) $24x^2 - 49x - 40$
15) $24x^2 + 8x - 2$
16) $25x^2 + 25x - 36$

17) $(4x + y)$
18) $8x^2 - 22x - 6$
19) $16x^2 + 24x + 9$
20) $40x$
21) $x^3 + 6x^2 + 12x + 8$
22) $(8x + 2y)$

Chapter 3:
Radicals Expressions

Topics that you will practice in this chapter:

- ✓ Square Roots
- ✓ Simplifying Radical Expressions
- ✓ Simplifying Radical Expressions Involving Fractions
- ✓ Multiplying Radical Expressions
- ✓ Adding and Subtracting Radical Expressions
- ✓ Domain and Range of Radical Functions
- ✓ Solving Radical Equations

Mathematics is no more computation than typing is literature.
– John Allen Paulos

Square Roots

✎ **Find the value each square root.**

1) $\sqrt{64} = \underline{}$

2) $\sqrt{4} = \underline{}$

3) $\sqrt{289} = \underline{}$

4) $\sqrt{0.25} = \underline{}$

5) $\sqrt{0.01} = \underline{}$

6) $\sqrt{0.09} = \underline{}$

7) $\sqrt{1,600} = \underline{}$

8) $\sqrt{2.25} = \underline{}$

9) $\sqrt{0} = \underline{}$

10) $\sqrt{0.04} = \underline{}$

11) $\sqrt{0.36} = \underline{}$

12) $\sqrt{0.81} = \underline{}$

13) $\sqrt{0.49} = \underline{}$

14) $\sqrt{1.21} = \underline{}$

15) $\sqrt{1.69} = \underline{}$

16) $\sqrt{0.16} = \underline{}$

17) $\sqrt{529} = \underline{}$

18) $\sqrt{625} = \underline{}$

19) $\sqrt{0.81} = \underline{}$

20) $\sqrt{20} = \underline{}$

21) $\sqrt{50} = \underline{}$

22) $\sqrt{676} = \underline{}$

23) $\sqrt{270} = \underline{}$

24) $\sqrt{32} = \underline{}$

✎ **Evaluate.**

25) $\sqrt{4} \times \sqrt{16} = \underline{}$

26) $\sqrt{49} \times \sqrt{64} = \underline{}$

27) $\sqrt{2} \times \sqrt{8} = \underline{}$

28) $\sqrt{17} \times \sqrt{17} = \underline{}$

29) $\sqrt{13} \times \sqrt{13} = \underline{}$

30) $\sqrt{15} \times \sqrt{15} = \underline{}$

31) $\sqrt{19} + \sqrt{19} = \underline{}$

32) $\sqrt{1} + \sqrt{1} = \underline{}$

33) $8\sqrt{7} - 2\sqrt{7} = \underline{}$

34) $7\sqrt{10} \times 6\sqrt{10} = \underline{}$

35) $9\sqrt{5} \times 2\sqrt{5} = \underline{}$

36) $8\sqrt{3} - \sqrt{12} = \underline{}$

Simplifying Radical Expressions

✎ **Simplify.**

1) $\sqrt{13y^2} =$

2) $\sqrt{60x^3} =$

3) $\sqrt[3]{27a} =$

4) $\sqrt{81x^2} =$

5) $\sqrt{150a} =$

6) $\sqrt[3]{135w^3} =$

7) $\sqrt{200x} =$

8) $\sqrt{192v} =$

9) $\sqrt[3]{64x} =$

10) $\sqrt{84x^3} =$

11) $\sqrt{121x^2} =$

12) $\sqrt[3]{48a} =$

13) $\sqrt{480} =$

14) $\sqrt{1,575p^2} =$

15) $\sqrt{108m^6} =$

16) $\sqrt{198x^3y^2} =$

17) $\sqrt{169x^2y^3} =$

18) $\sqrt{25a^6} =$

19) $\sqrt{50x^2y^3} =$

20) $\sqrt[3]{512y^3} =$

21) $2\sqrt{144x^2} =$

22) $3\sqrt{400x^2} =$

23) $\sqrt[3]{189xy^4} =$

24) $\sqrt[3]{1,331x^3y^5} =$

25) $3\sqrt{150a} =$

26) $\sqrt[3]{729y} =$

27) $3\sqrt{18xyr^3} =$

28) $6\sqrt{225x^2yz^6} =$

29) $3\sqrt[3]{125x^3y^2} =$

30) $7\sqrt{12a^2bc^4} =$

31) $4\sqrt[3]{1,000x^9y^{15}} =$

Algebra 2 Workbook

Multiplying Radical Expressions

✎ **Simplify.**

1) $\sqrt{11} \times \sqrt{11} =$

2) $\sqrt{5} \times \sqrt{15} =$

3) $\sqrt{3} \times \sqrt{12} =$

4) $\sqrt{20} \times \sqrt{25} =$

5) $\sqrt{5} \times (-2)\sqrt{35} =$

6) $2\sqrt{12} \times \sqrt{3} =$

7) $4\sqrt{24} \times \sqrt{6} =$

8) $\sqrt{5} \times (-\sqrt{75}) =$

9) $\sqrt{88} \times \sqrt{40} =$

10) $2\sqrt{45} \times 4\sqrt{105} =$

11) $\sqrt{32}(2 + \sqrt{2}) =$

12) $\sqrt{13x^2} \times \sqrt{13x} =$

13) $-7\sqrt{12} \times \sqrt{3} =$

14) $5\sqrt{19x^3} \times \sqrt{19x^3} =$

15) $\sqrt{15x^2} \times \sqrt{5x} =$

16) $-8\sqrt{2x} \times \sqrt{6x^5} =$

17) $-4\sqrt{5x} \times 5\sqrt{45x^2} =$

18) $-3\sqrt{27}(3 + \sqrt{15}) =$

19) $\sqrt{8x}(3 - \sqrt{2x}) =$

20) $\sqrt{5x}(10\sqrt{5x} + \sqrt{40}) =$

21) $\sqrt{18r}(6 + \sqrt{6}) =$

22) $-12\sqrt{3x} \times 3\sqrt{15x^3} =$

23) $-4\sqrt{27x} \times 6\sqrt{3x}$

24) $-3\sqrt{10v^2}(-3\sqrt{15v}) =$

25) $(\sqrt{8} - 3)(\sqrt{8} + \sqrt{9}) =$

26) $(-3\sqrt{5} + 7)(\sqrt{5} - 3) =$

27) $(3 - 4\sqrt{5})(-2 + \sqrt{4}) =$

28) $(13 - 2\sqrt{5})(3 - \sqrt{5}) =$

29) $(5 - \sqrt{3x})(5 + \sqrt{3x}) =$

30) $(-6 + 3\sqrt{3r})(-6 + \sqrt{3r}) =$

31) $(-\sqrt{5n} + 8)(-\sqrt{5} - 8) =$

32) $(-3 + 3\sqrt{2})(3 - 2\sqrt{2x}) =$

WWW.MathNotion.Com

Simplifying Radical Expressions Involving Fractions

✎ **Simplify.**

1) $\dfrac{\sqrt{3}}{\sqrt{2}} =$

2) $\dfrac{\sqrt{24}}{\sqrt{40}} =$

3) $\dfrac{\sqrt{12}}{2\sqrt{6}} =$

4) $\dfrac{21}{\sqrt{5}} =$

5) $\dfrac{15\sqrt{8r}}{\sqrt{m^5}} =$

6) $\dfrac{8\sqrt{2}}{\sqrt{m}} =$

7) $\dfrac{15\sqrt{25n^2}}{5\sqrt{15n}} =$

8) $\dfrac{\sqrt{8x^3y^5}}{\sqrt{2y^2x^4}} =$

9) $\dfrac{2}{2+\sqrt{5}} =$

10) $\dfrac{2-12\sqrt{x}}{\sqrt{24x}} =$

11) $\dfrac{2\sqrt{x}}{\sqrt{x}-\sqrt{y}} =$

12) $\dfrac{3-\sqrt{5}}{5-\sqrt{3}} =$

13) $\dfrac{5+\sqrt{12}}{5-\sqrt{3}} =$

14) $\dfrac{8}{-3-3\sqrt{3}} =$

15) $\dfrac{5}{2+\sqrt{15}} =$

16) $\dfrac{\sqrt{11}-\sqrt{7}}{\sqrt{7}-\sqrt{11}} =$

17) $\dfrac{\sqrt{8}+\sqrt{2}}{\sqrt{2}-\sqrt{8}} =$

18) $\dfrac{2\sqrt{2}-\sqrt{3}}{3\sqrt{2}+\sqrt{3}} =$

19) $\dfrac{\sqrt{11}+3\sqrt{5}}{3-\sqrt{11}} =$

20) $\dfrac{\sqrt{7}+\sqrt{13}}{13-\sqrt{7}} =$

21) $\dfrac{\sqrt{125a^7b^5}}{\sqrt{5ab^4}} =$

22) $\dfrac{72\sqrt{24m^3}}{9\sqrt{m}} =$

Adding and Subtracting Radical Expressions

✎ **Simplify.**

1) $\sqrt{5} + \sqrt{20} =$

2) $7\sqrt{44} + 7\sqrt{11} =$

3) $9\sqrt{3} - 3\sqrt{12} =$

4) $8\sqrt{9} - 5\sqrt{3} =$

5) $9\sqrt{80} - 9\sqrt{20} =$

6) $-\sqrt{32} - 5\sqrt{8} =$

7) $-12\sqrt{16} - 9\sqrt{64} =$

8) $15\sqrt{8} + 6\sqrt{32} =$

9) $16\sqrt{9} - 12\sqrt{36} =$

10) $-7\sqrt{7} + 11\sqrt{63} =$

11) $-24\sqrt{13} + 18\sqrt{117} =$

12) $25\sqrt{5} - 12\sqrt{45} =$

13) $-8\sqrt{99} + 2\sqrt{11} =$

14) $6\sqrt{3} - 2\sqrt{12} =$

15) $8\sqrt{20} + 3\sqrt{5} =$

16) $5\sqrt{28} - 8\sqrt{63} =$

17) $\sqrt{144} - \sqrt{121} =$

18) $6\sqrt{18} - 2\sqrt{2} =$

19) $-12\sqrt{7} + 21\sqrt{28} =$

20) $5\sqrt{60} - 5\sqrt{15} =$

21) $6\sqrt{54} - 3\sqrt{6} =$

22) $-4\sqrt{3} + 8\sqrt{75} =$

23) $-9\sqrt{20} - 7\sqrt{5} =$

24) $-\sqrt{216x} + 6\sqrt{6x} =$

25) $\sqrt{14y^2} + y\sqrt{126} =$

26) $\sqrt{11mn^2} + n\sqrt{99m} =$

27) $-8\sqrt{48a} - 2\sqrt{3a} =$

28) $-15\sqrt{17ab} - 10\sqrt{68ab} =$

29) $\sqrt{92x^2y} + x\sqrt{23y} =$

30) $5\sqrt{5a} + 4\sqrt{80a} =$

Answers of Worksheets – Chapter 3

Square Roots

1) 8
2) 2
3) 17
4) 0.5
5) 0.1
6) 0.3
7) 40
8) 1.5
9) 0
10) 0.2
11) 0.6
12) 0.9
13) 0.7
14) 1.1
15) 1.3
16) 0.4
17) 23
18) 25
19) 0.9
20) $2\sqrt{5}$
21) $5\sqrt{2}$
22) 26
23) $3\sqrt{30}$
24) $4\sqrt{2}$
25) 8
26) 56
27) 4
28) 17
29) 13
30) 15
31) $2\sqrt{19}$
32) 2
33) $6\sqrt{7}$
34) 420
35) 90
36) $6\sqrt{3}$

Simplifying radical expressions

1) $y\sqrt{13}$
2) $2x\sqrt{15x}$
3) $3\sqrt[3]{a}$
4) $9x$
5) $5\sqrt{6a}$
6) $3w\sqrt[3]{5}$
7) $10\sqrt{2x}$
8) $8\sqrt{3v}$
9) $4\sqrt[3]{x}$
10) $2x\sqrt{21x}$
11) $11x$
12) $2\sqrt[3]{6a}$
13) $4\sqrt{30}$
14) $15p\sqrt{7}$
15) $6m^3\sqrt{3}$
16) $3x.y\sqrt{22x}$
17) $13xy\sqrt{y}$
18) $5a^3$
19) $5xy\sqrt{2y}$
20) $8y$
21) $24x$
22) $60x$
23) $3y\sqrt[3]{7xy}$
24) $11xy\sqrt[3]{y^2}$
25) $15\sqrt{6a}$
26) $9\sqrt[3]{y}$
27) $9r\sqrt{2xyr}$
28) $90xz^3\sqrt{y}$
29) $15x\sqrt[3]{y^2}$
30) $14ac^2\sqrt{b}$
31) $40x^3y^{15}$

Multiplying radical expressions

1) 11
2) $5\sqrt{3}$
3) 6
4) $10\sqrt{5}$
5) $-10\sqrt{7}$
6) 12
7) 48
8) $-5\sqrt{15}$
9) $8\sqrt{55}$
10) $120\sqrt{21}$
11) $8\sqrt{2}+8$
12) $13x\sqrt{x}$

Algebra 2 Workbook

13) -42
14) $95x^3$
15) $5x\sqrt{3x}$
16) $-16x^3\sqrt{3}$
17) $-300x\sqrt{x}$
18) $-27\sqrt{3} - 27\sqrt{5}$
19) $6\sqrt{2x} - 4x$

20) $50x + 2\sqrt{50x}$
21) $18\sqrt{2r} + 6\sqrt{3r}$
22) $-108x^2\sqrt{5}$
23) $-216x$
24) $47v\sqrt{6v}$
25) -1
26) $16\sqrt{5} - 36$

27) 0
28) $49 - 19\sqrt{5}$
29) $28 - 3x$
30) $9r - 24\sqrt{3r} + 36$
31) $5\sqrt{n} - 64$
32) $-9 + 6\sqrt{2x} + 9\sqrt{2} - 12\sqrt{x}$

Simplifying radical expressions involving fractions

1) $\frac{\sqrt{6}}{2}$
2) $\frac{\sqrt{15}}{5}$
3) $\frac{\sqrt{2}}{2}$
4) $\frac{21\sqrt{5}}{5}$
5) $\frac{30\sqrt{2mr}}{m^3}$
6) $\frac{8\sqrt{2m}}{m}$
7) $\sqrt{15n}$

8) $\frac{2y\sqrt{xy}}{x}$
9) $-2(2 - \sqrt{2})$
10) $\frac{\sqrt{6x}\,(1-6\sqrt{x})}{6x}$
11) $\frac{2\sqrt{x}\,(\sqrt{x}+\sqrt{y})}{x-y}$
12) $\frac{15+ 3\sqrt{3}- 5\sqrt{5}-\sqrt{15}}{22}$
13) $\frac{31+15\sqrt{3}}{22}$
14) $-\frac{4(\sqrt{3}-1)}{3}$

15) $-\frac{5(2-\sqrt{15})}{11}$
16) -1
17) -3
18) $\frac{3-\sqrt{6}}{3}$
19) $-\frac{3\sqrt{11}+11+9\sqrt{5}+3\sqrt{55}}{2}$
20) $\frac{13\sqrt{7}+7+13\sqrt{13}+\sqrt{91}}{162}$
21) $5a^3\sqrt{b}$
22) $16\sqrt{6}\,m$

Adding and subtracting radical expressions

1) $2\sqrt{5}$
2) $21\sqrt{11}$
3) $-5\sqrt{3}$
4) $24 - 5\sqrt{3}$
5) $18\sqrt{5}$
6) $-14\sqrt{2}$
7) -120
8) $54\sqrt{2}$
9) -24
10) $26\sqrt{7}$

11) $30\sqrt{13}$
12) $-11\sqrt{5}$
13) $-22\sqrt{11}$
14) $2\sqrt{3}$
15) $19\sqrt{5}$
16) $-14\sqrt{7}$
17) 1
18) $16\sqrt{2}$
19) $30\sqrt{7}$
20) $5\sqrt{15}$

21) $15\sqrt{6}$
22) $36\sqrt{3}$
23) $-25\sqrt{5}$
24) 0
25) $4y\sqrt{14}$
26) $4n\sqrt{11m}$
27) $-34\sqrt{3a}$
28) $-35\sqrt{17ab}$
29) $-x\sqrt{23y}$
30) $21\sqrt{5a}$

Chapter 4: Functions Operations and Quadratic

Topics that you will practice in this chapter:

- ✓ Evaluating Functions
- ✓ Adding and Subtracting Functions
- ✓ Multiply and Dividing Functions
- ✓ Composition of Functions
- ✓ Solving Quadratic Equations
- ✓ Quadratic Formula and Discriminant
- ✓ Quadratic Inequalities
- ✓ Graphing Quadratic Functions
- ✓ Domain and Range of Radical Functions
- ✓ Solving Radical Equations

It's fine to work on any problem, so long as it generates interesting mathematics along the way – even if you don't solve it at the end of the day." – Andrew Wiles

Evaluating Function

✍ **Write each of following in function notation.**

1) $h = -8x + 9$

2) $k = 5a - 21$

3) $d = 14t$

4) $y = \frac{3}{17}x - \frac{9}{17}$

5) $m = 18n - 94$

6) $c = p^2 - 7p + 15$

✍ **Evaluate each function.**

7) $f(x) = 6x - 7$, find $f(-3)$

8) $g(x) = \frac{1}{10}x + 6$, find $f(5)$

9) $h(x) = -2x + 15$, find $f(8)$

10) $f(x) = -3x + 8$, find $f(-2)$

11) $f(a) = 12a - 9$, find $f(0)$

12) $h(x) = 18 - 5x$, find $f(-4)$

13) $g(n) = 7n - 5$, find $f(5)$

14) $f(x) = -9x - 2$, find $f(3)$

15) $k(n) = -12 + 4.5n$, find $f(2)$

16) $f(x) = -1.5x + 2.5$, find $f(-6)$

17) $g(n) = \frac{16n-8}{6n}$, find $g(2)$

18) $g(n) = \sqrt{5n} - 2$, find $g(5)$

19) $h(x) = x^{-1} - 6$, find $h(\frac{1}{9})$

20) $h(n) = n^{-3} + 4$, find $h(\frac{1}{2})$

21) $h(n) = n^2 - 5$, find $h(\frac{4}{5})$

22) $h(n) = n^3 - 8$, find $h(-\frac{1}{3})$

23) $h(n) = 4n^2 - 42$, find $h(-4)$

24) $h(n) = -5n^2 - 9n$, find $h(7)$

25) $g(n) = \sqrt{4n^2} - \sqrt{5n}$, find $g(5)$

26) $h(a) = \frac{-15a+7}{3a}$, find $h(-b)$

27) $k(a) = 8a - 9$, find $k(a - 3)$

28) $h(x) = \frac{1}{6}x + 7$, find $h(-12x)$

29) $h(x) = 8x^2 + 10$, find $h(\frac{x}{2})$

30) $h(x) = x^4 - 8$, find $h(-2x)$

Algebra 2 Workbook

Adding and Subtracting Functions

✎ **Perform the indicated operation.**

1) $f(x) = 2x + 3$

   $g(x) = x + 4$

   Find $(f - g)(2)$

2) $g(a) = -3a - 8$

   $f(a) = -4a - 12$

   Find $(g - f)(-2)$

3) $h(t) = 7t + 5$

   $g(t) = 3t + 11$

   Find $(h - g)(t)$

4) $g(a) = -5a - 3$

   $f(a) = 3a^2 + 4$

   Find $(g - f)(x)$

5) $g(x) = \frac{2}{7}x - 10$

   $h(x) = \frac{5}{7}x + 10$

   Find $g(14) - h(14)$

6) $h(3) = \sqrt{7x} - 2$

   $g(x) = \sqrt{7x} + 2$

   Find $(h + g)(7)$

7) $f(x) = x^{-3}$

   $g(x) = x^2 + \frac{4}{x}$

   Find $(f - g)(-1)$

8) $h(n) = n^2 + 8$

   $g(n) = -n + 5$

   Find $(h - g)(a)$

9) $g(x) = -2x^2 - 3 - x$

   $f(x) = 7 + x$

   Find $(g - f)(2x)$

10) $g(t) = 4t - 9$

    $f(t) = -t^2 + 5$

    Find $(g + f)(-z)$

11) $f(x) = 3x + 9$

    $g(x) = -4x^2 + 2x$

    Find $(f - g)(-x^2)$

12) $f(x) = -9x^3 - 4x$

    $g(x) = 4x + 12$

    Find $(f + g)(3x^2)$

www.MathNotion.Com

Multiplying and Dividing Functions

✏️ **Perform the indicated operation.**

1) $g(x) = -2x - 5$
   $f(x) = 3x + 4$
   Find $(g \cdot f)(2)$

2) $f(x) = 3x$
   $h(x) = -2x + 5$
   Find $(f \cdot h)(-3)$

3) $g(a) = 5a - 3$
   $h(a) = a - 7$
   Find $(g \cdot h)(-3)$

4) $f(x) = x - 4$
   $h(x) = 4x - 3$
   Find $\left(\frac{f}{h}\right)(4)$

5) $f(x) = 9a^2$
   $g(x) = 5 + 4a$
   Find $\left(\frac{f}{g}\right)(3)$

6) $g(a) = \sqrt{5a} + 7$
   $f(a) = (-a)^2 + 3$
   Find $\left(\frac{g}{f}\right)(5)$

7) $g(t) = t^2 + 5$
   $h(t) = 2t - 5$
   Find $(g \cdot h)(-3)$

8) $g(n) = n^2 + 2n - 4$
   $h(n) = -n + 6$
   Find $(g \cdot h)(1)$

9) $g(a) = (a - 7)^3$
   $f(a) = a^2 + 8$
   Find $\left(\frac{g}{f}\right)(7)$

10) $g(x) = -x^2 + \frac{4}{5}x + 10$
    $f(x) = x^2 - 3$
    Find $\left(\frac{g}{f}\right)(5)$

11) $f(x) = x^3 - 3x^2 + 9$
    $g(x) = x - 4$
    Find $(f \cdot g)(x)$

12) $f(x) = 3x - 5$
    $g(x) = x^2 - 4x$
    Find $(f \cdot g)(x^2)$

Composition of Functions

Using $f(x) = 2x - 5$ and $g(x) = -2x$, find:

1) $f(g(0)) =$

2) $f(g(-1)) =$

3) $g(f(1)) =$

4) $g(f(3)) =$

5) $f(g(-2)) =$

6) $g(f(5)) =$

Using $f(x) = -\frac{1}{4}x + \frac{3}{4}$ and $g(x) = x^2$, find:

7) $g(f(4)) =$

8) $g(f(3)) =$

9) $g(g(2)) =$

10) $f(f(1)) =$

11) $g(f(-1)) =$

12) $g(f(7)) =$

Using $f(x) = -5x + 2$ and $g(x) = x + 3$, find:

13) $g(f(0)) =$

14) $f(f(2)) =$

15) $f(g(3)) =$

16) $f(g(-3)) =$

17) $g(f(-5)) =$

18) $f(f(x)) =$

Using $f(x) = \sqrt{x + 9}$ and $g(x) = x - 9$, find:

19) $f(g(9)) =$

20) $g(f(-8)) =$

21) $f(g(18)) =$

22) $f(f(-5)) =$

23) $g(f(7)) =$

24) $g(g(8)) =$

Quadratic Equation

🔖 **Multiply.**

1) $(x-2)(x+8) = $ _____

2) $(x+1)(x+9) = $ _____

3) $(x-5)(x+6) = $ _____

4) $(x+7)(x-3) = $ _____

5) $(x-9)(x-8) = $ _____

6) $(4x+2)(x-4) = $ _____

7) $(3x-6)(x+4) = $ _____

8) $(x-9)(2x+7) = $ _____

9) $(5x+3)(x-4) = $ _____

10) $(4x+2)(3x-3) = $ _____

🔖 **Factor each expression.**

11) $x^2 - 4x - 21 = $ _____

12) $x^2 + 14x + 45 = $ _____

13) $x^2 - 5x - 24 = $ _____

14) $x^2 - 7x + 6 = $ _____

15) $x^2 + 14x + 33 = $ _____

16) $4x^2 + 38x + 18 = $ _____

17) $5x^2 + 18x - 8 = $ _____

18) $2x^2 + 2x - 40 = $ _____

19) $2x^2 + 22x + 56 = $ _____

20) $12x^2 - 148x + 360 = $ _____

🔖 **Calculate each equation.**

21) $(x+6)(x-9) = 0$

22) $(x+1)(x+11) = 0$

23) $(3x+9)(x+3) = 0$

24) $(5x-5)(6x+12) = 0$

25) $x^2 - 12x + 30 = 6$

26) $x^2 + 6x + 14 = 5$

27) $x^2 + \frac{9}{2}x + 7 = 5$

28) $x^2 + 2x - 25 = 10$

29) $2x^2 + 12x - 54 = 0$

30) $x^2 - 11x = 12$

Solving Quadratic Equations

✎ **Solve each equation by factoring or using the quadratic formula.**

1) $(x+5)(x-2)=0$

2) $(x+8)(x+2)=0$

3) $(x-9)(x+5)=0$

4) $(x-3)(x-1)=0$

5) $(x+9)(x+4)=0$

6) $(2x+5)(x+9)=0$

7) $(9x+8)(3x+9)=0$

8) $(4x+2)(x+5)=0$

9) $(x+2)(2x+9)=0$

10) $(12x+3)(2x+9)=0$

11) $2x^2=16x$

12) $x^2-16=0$

13) $2x^2+48=22x$

14) $-x^2-20=9x$

15) $x^2+8x=33$

16) $2x^2+12x=80$

17) $x^2+14x=-48$

18) $x^2+15x=-54$

19) $x^2+15x=-36$

20) $x^2+2x-40=5x$

21) $x^2+16x=-63$

22) $x^2-18x=-81$

23) $10x^2=7x-1$

24) $7x^2-5x+8=8$

25) $8x^2+27=33x$

26) $5x^2-26x=-24$

27) $3x^2+6=-19x$

28) $x^2+22x=-117$

29) $x^2+3x-58=30$

30) $5x^2+20x-200=25$

31) $3x^2-33x+84=0$

32) $6x^2-31x+30=15-10x^2$

Algebra 2 Workbook

Quadratic Formula and the Discriminant

✎ **Find the value of the discriminant of each quadratic equation.**

1) $3x(x - 9) = 0$

2) $2x^2 + 9x - 4 = 0$

3) $x^2 + 9x + 5 = 0$

4) $4x^2 - 4x + 7 = 0$

5) $x^2 + 7x - 6 = 0$

6) $4x^2 + 5x - 13 = 0$

7) $3x^2 + 7x + 11 = 0$

8) $x^2 - 4x - 12 = 0$

9) $5x^2 + 9x + 8 = 0$

10) $x^2 + 3x - 7 = 0$

11) $6x^2 + 7x - 13 = 0$

12) $-8x^2 - 11x + 9 = 0$

13) $-9x^2 - 13x + 7 = 0$

14) $-6x^2 - 7x - 9 = 0$

15) $14x^2 - 8x - 15 = 0$

16) $-9x^2 - 5x + 10 = 0$

17) $8x^2 + 9x - 14 = 0$

18) $7x^2 - 15x = 0$

19) $3x^2 - 7x + 9 = 0$

20) $7x^2 + 4x + 16 = 0$

✎ **Find the discriminant of each quadratic equation then state the number of real and imaginary solutions.**

21) $-x^2 - 4 = 4x$

22) $20x^2 = 20x - 5$

23) $-11x^2 - 11x = 22$

24) $19x^2 - 4x + 1 = 15x^2$

25) $-8x^2 = -6x + 6$

26) $2x^2 + 4x + 4 = 2$

27) $6x^2 - 2x - 9 = -12$

28) $-14x^2 - 56x - 64 = -8$

WWW.MathNotion.Com

Quadratic Inequalities

✎ **Solve each quadratic inequality.**

1) $x^2 - 64 < 0$

2) $-x^2 - 6x - 8 > 0$

3) $x^2 + 6x + 8 < 0$

4) $4x^2 + 28x + 40 > 0$

5) $5x^2 - 5x - 10 \geq 0$

6) $3x^2 > -12x - 27$

7) $4x^2 + 10x + 28 \leq 0$

8) $3x^2 - 9x - 30 \leq 0$

9) $5x^2 - 35x + 60 \geq 0$

10) $x^2 + 7x + 10 < 0$

11) $2x^2 + 16x - 130 > 0$

12) $8x^2 - 24x + 18 > 0$

13) $2x^2 - 32x + 136 \leq 0$

14) $x^2 - 14x + 49 \leq 0$

15) $2x^2 - 30x + 112 \geq 0$

16) $2x^2 + 16x + 32 \leq 0$

17) $x^2 - 121 \leq 0$

18) $2x^2 - 22x + 60 \geq 0$

19) $8x^2 + 10x + 18 \leq 0$

20) $4x^2 - 2x - 24 > 2x^2$

21) $4x^2 - 16x + 16 < 0$

22) $15x^2 - 6x \geq 14x^2 - 5$

23) $6x^2 - 24 > 4x^2 + 2x$

24) $3x^2 - x \geq 3x^2 - 4x + 6$

25) $2x^2 + 2x - 8 > x^2$

26) $4x^2 + 20x - 11 < 0$

27) $-2x^2 + 30x - 114 \geq 0$

28) $-8x^2 + 6x - 1 \leq 0$

29) $x^2 + 7x + 10 < 0$

30) $36x^2 + 46x + 10 \leq 0$

31) $5x^2 + 5x - 60 \geq 0$

32) $3x^2 + 4x \leq 2x^2 + 2x - 10$

WWW.MathNotion.Com

Graphing Quadratic Functions

📝 **Sketch the graph of each function. Identify the vertex and axis of symmetry.**

1) $y = (x + 3)^2 + 5$

2) $y = (x - 3)^2 - 1$

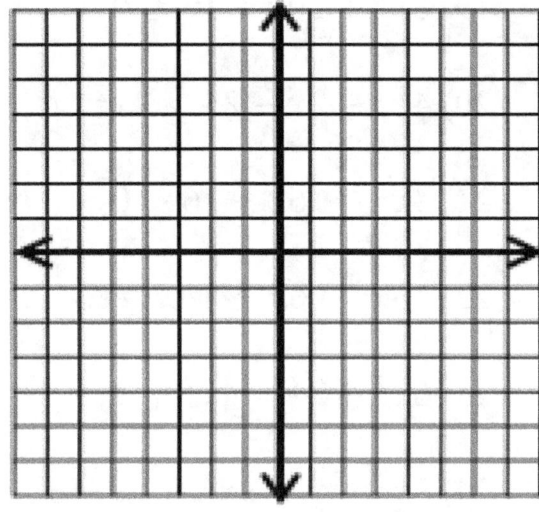 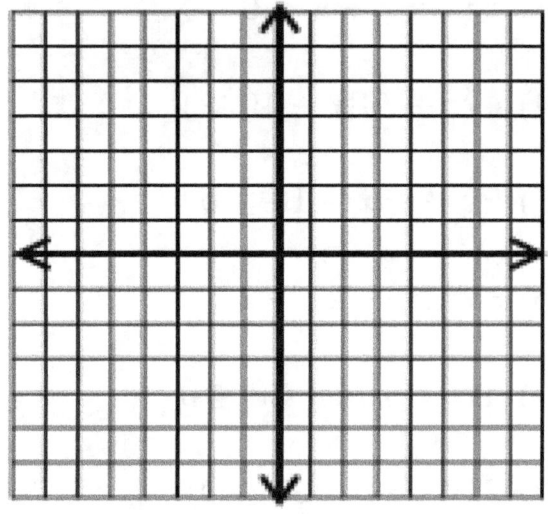

3) $y = 6 - (-x + 2)^2$

4) $y = -3x^2 - 6x + 9$

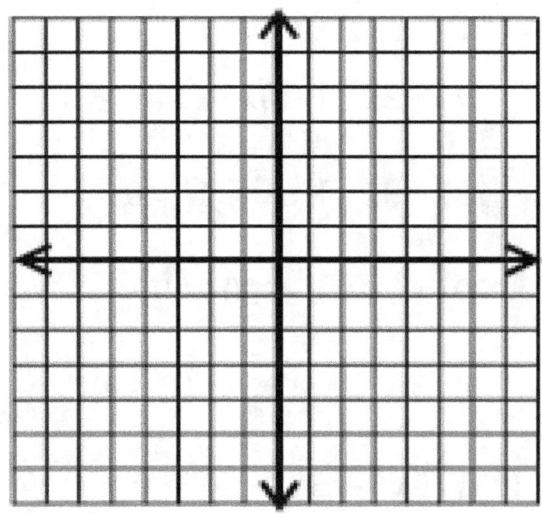 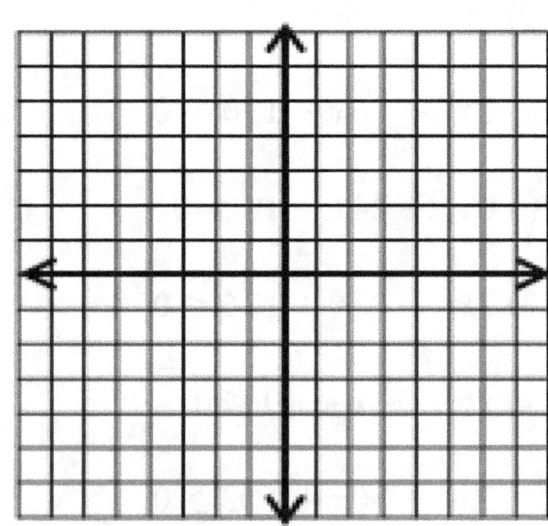

Domain and Range of Radical Functions

✍ **Identify the domain and range of each function.**

1) $y = \sqrt{x+6} - 9$

2) $y = \sqrt[3]{5x-4} - 12$

3) $y = \sqrt{3x-9} + 7$

4) $y = \sqrt[3]{(8x+11)} - 9$

5) $y = 2\sqrt{6x+30} + 14$

6) $y = \sqrt[3]{(9x-15)} - 17$

7) $y = 2\sqrt{9x^2+18} + 7$

8) $y = \sqrt[3]{(8x^2-5)} - 13$

9) $y = \sqrt{2x^3+16} - 9$

10) $y = \sqrt[3]{(14x+3)} - 5x$

11) $y = 3\sqrt{-3(12x+24)} + 7$

12) $y = \sqrt[5]{(11x^2-17)} - 21$

13) $y = 4\sqrt{x-9} - 27$

14) $y = \sqrt[3]{5x+1} - 3$

✍ **Sketch the graph of each function.**

15) $y = -3\sqrt{x} + 4$

16) $y = 6\sqrt{x} - 8$

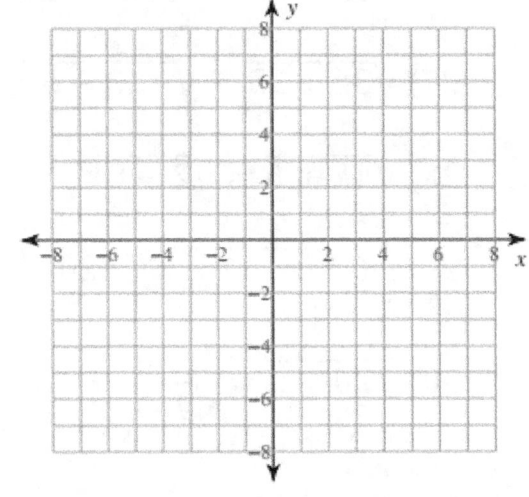
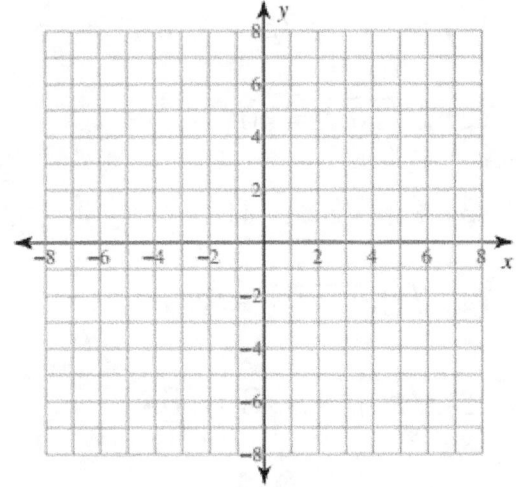

Algebra 2 Workbook

Solving Radical Equations

✎ **Solve each equation. Remember to check for extraneous solutions.**

1) $\sqrt{a} = 9$

2) $\sqrt{v} = 4$

3) $\sqrt{r} = 7$

4) $4 = 16\sqrt{x}$

5) $\sqrt{x+3} = 12$

6) $2 = \sqrt{x-8}$

7) $7 = \sqrt{r-6}$

8) $\sqrt{x-4} = 9$

9) $15 = \sqrt{x-6}$

10) $\sqrt{m+8} = 11$

11) $5\sqrt{3a} = 75$

12) $2\sqrt{10x} = 30$

13) $4 = \sqrt{3x-10}$

14) $\sqrt{150-3x} = 2$

15) $\sqrt{r+4} - 8 = 8$

16) $-18 = -3\sqrt{r+2}$

17) $60 = 6\sqrt{49v}$

18) $3 = \sqrt{50-x}$

19) $\sqrt{90-5a} = 6$

20) $\sqrt{-3n+33} = 3$

21) $\sqrt{21r-18} = 4r$

22) $\sqrt{-14+6x} = 7x$

23) $\sqrt{4x+15} = \sqrt{2x+11}$

24) $\sqrt{8v} = \sqrt{10v-14}$

25) $\sqrt{16-3x} = \sqrt{3x-8}$

26) $\sqrt{5m+12} = \sqrt{7m+12}$

27) $\sqrt{8r+15} = \sqrt{-13-5r}$

28) $\sqrt{4k+6} = \sqrt{2-8k}$

29) $-60\sqrt{x-16} = -120$

30) $\sqrt{20-x} = \sqrt{\dfrac{x}{4}}$

Algebra 2 Workbook

Answers of Worksheets – Chapter 4

Evaluating Function

1) $h(x) = -8x + 9$
2) $k(a) = 5a - 21$
3) $d(t) = 14t$
4) $y(x) = \frac{3}{17}x - \frac{9}{17}$
5) $m(n) = 18n - 94$
6) $c(p) = p^2 - 7p + 15$
7) -25
8) 6.5
9) -1
10) 14
11) -9
12) 38
13) 30
14) -29
15) -3
16) 11.5
17) 2
18) 3
19) 3
20) 12
21) $-\frac{109}{25}$
22) $-\frac{215}{27}$
23) 22
24) -308
25) 5
26) $-\frac{15b+7}{3b}$
27) $8a - 33$
28) $-2x + 7$
29) $2x^2 + 10$
30) $-16x^4 - 8$

Adding and Subtracting Functions

1) 1
2) 2
3) $4t - 6$
4) $-3x^2 - 5x - 7$
5) -26
6) 14
7) 2
8) $a^2 + a + 3$
9) $-8x^2 - 4x - 10$
10) $-z^2 - 4z - 4$
11) $4x^4 - x^2 + 9$
12) $-243x^6 + 12$

Multiplying and Dividing Functions

1) -90
2) -99
3) 180
4) 0
5) $\frac{81}{17}$
6) $\frac{3}{7}$
7) -154
8) -5
9) 0
10) $-\frac{1}{2}$
11) $x^4 - 7x^3 + 12x^2 + 9x - 36$
12) $3x^6 - 17x^4 + 20x^2$

Composition of Functions

1) -5
2) -1
3) 6
4) -2
5) 3
6) -10
7) $\frac{1}{16}$
8) 0

WWW.MathNotion.Com

Algebra 2 Workbook

9) 16	13) 5	17) 30	21) $3\sqrt{2}$
10) $\frac{5}{8}$	14) 42	18) $25x - 8$	22) $\sqrt{11}$
11) 1	15) -28	19) 3	23) -5
12) 1	16) 2	20) -8	24) -10

Quadratic Equations

1) $x^2 + 6x - 16$	11) $(x - 7)(x + 3)$	21) $x = -6, x = 9$
2) $x^2 + 10x + 9$	12) $(x + 5)(x + 9)$	22) $x = -1, x = -11$
3) $x^2 + x - 30$	13) $(x - 8)(x + 3)$	23) $x = -3$
4) $x^2 + 4x - 21$	14) $(x - 1)(x - 6)$	24) $x = 1, x = -2$
5) $x^2 - 17x + 72$	15) $(x + 3)(x + 11)$	25) $x = 6$
6) $4x^2 - 14x - 8$	16) $(4x + 2)(x + 9)$	26) $x = -3$
7) $3x^2 + 6x - 24$	17) $(5x - 2)(x + 4)$	27) $x = -4, x = -\frac{1}{2}$
8) $2x^2 - 11x - 63$	18) $(2x - 8)(x + 5)$	28) $x = 5, x = -7$
9) $5x^2 - 17x - 12$	19) $(2x + 8)(x + 7)$	29) $x = 3, x = -9$
10) $12x^2 - 6x - 6$	20) $4(x - 9)(3x - 10)$	30) $x = -1, x = 12$

Solving quadratic equations

1) $\{-5, 2\}$	11) $\{8, 0\}$	23) $\{\frac{1}{5}, \frac{1}{2}\}$
2) $\{-8, -2\}$	12) $\{4, -4\}$	24) $\{\frac{5}{7}, 0\}$
3) $\{9, -5\}$	13) $\{3, 8\}$	25) $\{\frac{9}{8}, 3\}$
4) $\{3, 1\}$	14) $\{-5, -4\}$	26) $\{\frac{6}{5}, 4\}$
5) $\{-9, -4\}$	15) $\{3, -11\}$	27) $\{-\frac{1}{3}, -6\}$
6) $\{-\frac{5}{2}, -9\}$	16) $\{4, -10\}$	28) $\{-9, -13\}$
7) $\{-\frac{8}{9}, -3\}$	17) $\{-6, -8\}$	29) $\{8, -11\}$
8) $\{-\frac{1}{2}, -5\}$	18) $\{-6, -9\}$	30) $\{5, -9\}$
9) $\{-2, -\frac{9}{2}\}$	19) $\{-3, -12\}$	31) $\{4, 7\}$
10) $\{-\frac{1}{4}, -\frac{9}{2}\}$	20) $\{8, -5\}$	32) $\{\frac{15}{16}, 1\}$
	21) $\{-7, -9\}$	
	22) $\{9\}$	

Quadratic formula and the discriminant

| 1) 729 | 2) 113 | 3) 61 |

Algebra 2 Workbook

4) -96
5) 73
6) 233
7) 83
8) 8
9) -79
10) 37
11) 361
12) 409
13) 421
14) 7
15) 904
16) 385
17) 529
18) 225
19) -59
20) -432
21) 0, one real solution
22) 0, one real solution
23) -847, no solution
24) 0, one real solution
25) -156, no solution
26) 0, one real solution
27) -68, no solution
28) 0, one real solution

Solve quadratic inequalities.

1) $-8 < x < 8$
2) $-4 < x < -2$
3) $-4 < x < -2$
4) $x < -5 \; or \; x > -2$
5) $x \leq -1 \; or \; x \geq 2$
6) all real numbers
7) no solution
8) $-2 \leq x \leq 5$
9) $x \leq 3 \; or \; x \geq 4$
10) $-5 < x < -2$
11) $x < -13 \; or \; x > 5$
12) $x < \frac{3}{2} \; or \; x > \frac{3}{2}$
13) no solution
14) $x = 7$
15) $x \leq 7 \; or \; x \geq 8$
16) $x = -4$
17) $-11 \leq x \leq 11$
18) $x \leq 5 \; or \; x \geq 6$
19) no solution
20) $x < -3 \; or \; x > 4$
21) no solution
22) $x \leq 1 \; or \; x \geq 5$
23) $x < -3 \; or \; x > 4$
24) $x \geq 2$
25) $x < -4 \; or \; x > 2$
26) $-\frac{11}{2} < x < \frac{1}{2}$
27) no solution
28) $x \leq \frac{1}{4} \; or \; x \geq \frac{1}{2}$
29) $-5 < x < -2$
30) $-1 \leq x \leq -\frac{5}{18}$
31) $x \leq -4 \; or \; x \geq 3$
32) no solution

Graphing quadratic functions

1) $(-3, 5), x = -3$

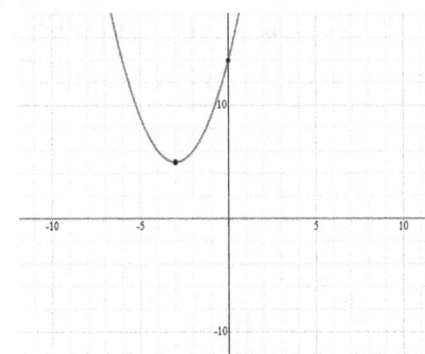

2) $(3, -1), x = 3$

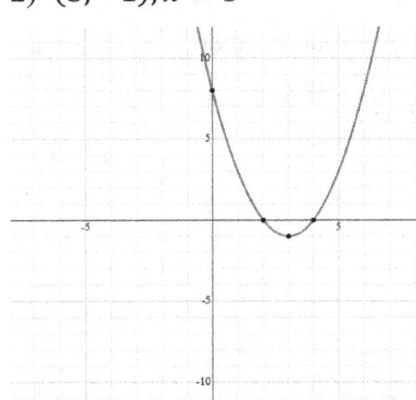

3) $(2, 6), x = 2$

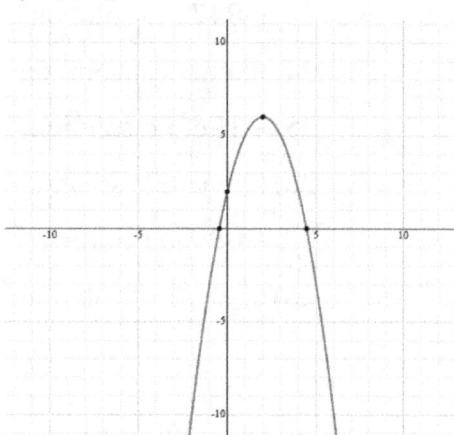

1) $(-1, 12), x = -1$

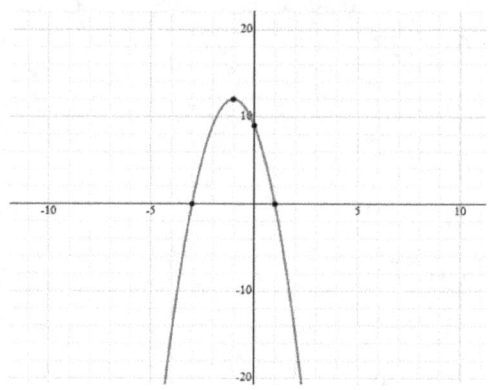

Domain and range of radical functions

1) domain: $x \geq -6$
   range: $y \geq -9$

2) domain: {all real numbers}
   range: {all real numbers}

3) domain: $x \geq 3$
   range: $y \geq 7$

4) domain: {all real numbers}
   range: {all real numbers}

5) domain: $x \geq -5$
   range: $y \geq 14$

6) domain: {all real numbers}
   range: {all real numbers}

7) domain: {all real numbers}
   range: $y \geq 6\sqrt{2} + 7$

8) domain: {all real numbers}
   range: {all real numbers}

9) domain: $x \geq -2$
   range: $y \geq -9$

10) domain: {all real numbers}
    range: {all real numbers}

11) domain: $x \leq -2$
    range: $y \geq 7$

12) domain: {all real numbers}
    range: {all real numbers}

13) domain: $x \geq 9$
    range: $y \geq -27$

14) domain: {all real numbers}
    range: {all real numbers}

15)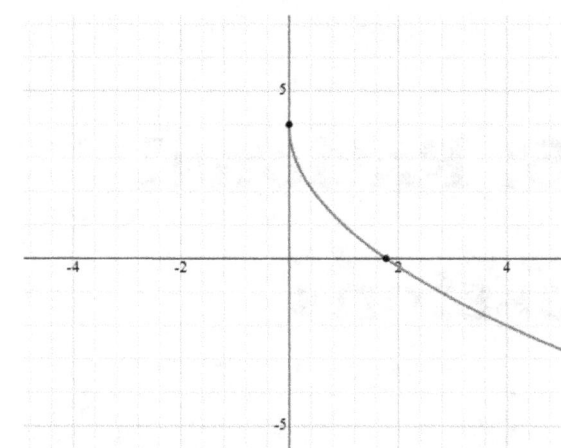

16)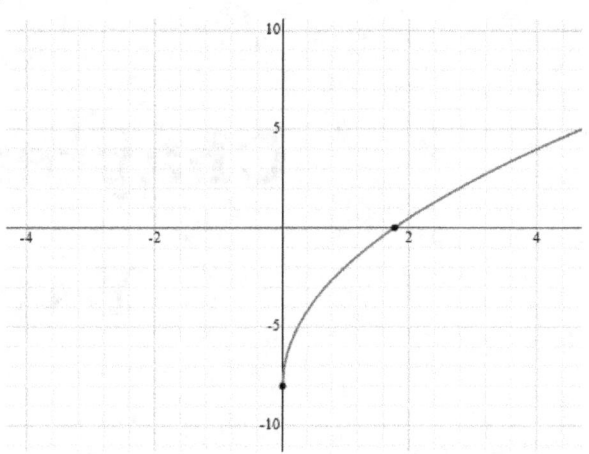

Solving radical equations

1) {81}
2) {16}
3) {49}
4) {$\frac{1}{16}$}
5) {141}
6) {12}
7) {55}
8) {85}
9) {231}
10) {113}

11) {75}
12) {22.5}
13) {$\frac{26}{3}$}
14) $\frac{146}{3}$
15) {252}
16) {34}
17) {$\frac{100}{49}$}
18) {41}
19) {54/5}
20) {8}

21) no solution
22) no solution
23) {−2}
24) {7}
25) {4}
26) {0}
27) no solution
28) {$-\frac{1}{3}$}
29) {20}
30) {16}

Chapter 5: Rational Expressions

Topics that you will learn in this chapter:

- ✓ Simplifying and Graphing Rational Expressions
- ✓ Adding and Subtracting Rational Expressions
- ✓ Multiplying and Dividing Rational Expressions
- ✓ Solving Rational Equations and Complex Fractions

"What music is to the heart, mathematics is to the mind."
— Amit Kalantri

Simplifying and Graphing Rational Expressions

✎ **Simplify.**

1) $\dfrac{x+2}{2x+4} =$

2) $\dfrac{2x^2+2x-12}{x+3} =$

3) $\dfrac{9}{3x-3} =$

4) $\dfrac{x^2-2x-3}{x^2+x-12} =$

5) $\dfrac{14x^3}{18x} =$

6) $\dfrac{x-2}{x^2+2x-8} =$

7) $\dfrac{x^2-6x-16}{x-8} =$

8) $\dfrac{25}{5x-5} =$

✎ **Identify the points of discontinuity, holes, vertical asymptotes, x-intercepts, and horizontal asymptote of each.**

9) $f(x) = \dfrac{x^3-x^2-6x}{-3x^3-3x+18} =$

10) $f(x) = \dfrac{x^2+x-6}{-4x^2-16x-12} =$

11) $f(x) = \dfrac{x-3}{x-9} =$

12) $f(x) = \dfrac{1}{2x^2-2x-12} =$

✎ **Graph rational expressions.**

13) $f(x) = \dfrac{x^2+2x-4}{x-2}$

14) $f(x) = \dfrac{4x^3-16x+64}{x^2-2x-4}$

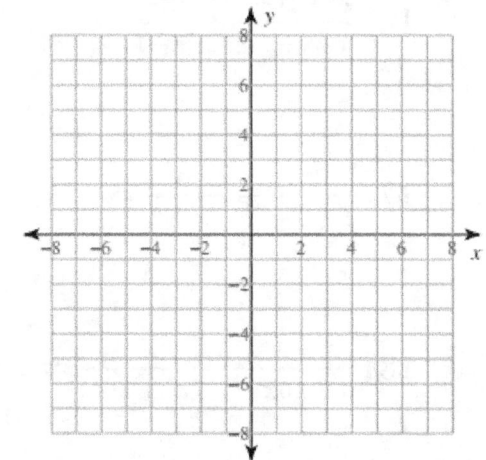
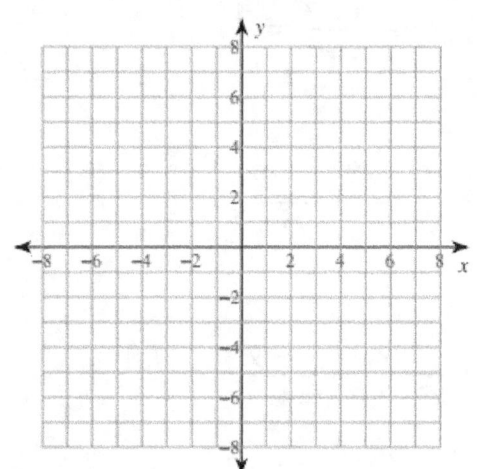

Adding and Subtracting Rational Expressions

✏️ Simplify each expression.

1) $\dfrac{2}{4x+10} + \dfrac{x-4}{4x+10} =$

2) $\dfrac{x+3}{x-4} + \dfrac{x-3}{x+3} =$

3) $\dfrac{3}{x+6} - \dfrac{4}{x-7} =$

4) $\dfrac{x-4}{x^2-12} - \dfrac{x-1}{12-x^2} =$

5) $\dfrac{3}{x+3} + \dfrac{4x}{2x+6} =$

6) $\dfrac{5+x}{x} + \dfrac{x-2}{x} =$

7) $2 + \dfrac{x-2}{x+1} =$

8) $\dfrac{2x}{5x+4} + \dfrac{4x}{2x+3} =$

9) $\dfrac{3xy}{x^2-y^2} - \dfrac{x-y}{x+y} =$

10) $\dfrac{2}{x^2-5x+6} + \dfrac{-2}{x^2-9} =$

11) $\dfrac{4}{x+1} - \dfrac{2}{x+3} =$

12) $\dfrac{5x+5}{5x^2+15x-20} + \dfrac{3x}{2x} =$

13) $3 + \dfrac{x}{x+3} - \dfrac{3}{x^2-9} =$

14) $\dfrac{2}{x+1} - \dfrac{2}{x+3} =$

15) $\dfrac{2}{3x^2+18x} + \dfrac{4}{x} =$

16) $\dfrac{x^2+2x+1}{2x+2} + \dfrac{2x-2}{x-1} =$

17) $\dfrac{2x}{5x+4} + \dfrac{8x}{2x+3} =$

18) $\dfrac{x+4}{5+20x} - \dfrac{x-4}{5x^2+20x} =$

Multiplying and Dividing Rational Expressions

🖎 **Simplify each expression.**

1) $\dfrac{14x}{12} \times \dfrac{12}{16x} =$

2) $\dfrac{78x}{25} \times \dfrac{95}{21x^2} =$

3) $\dfrac{78}{3} \times \dfrac{44x}{93} =$

4) $\dfrac{65}{43} \times \dfrac{43x^2}{37} =$

5) $\dfrac{93}{21x} \times \dfrac{34x}{51x} =$

6) $\dfrac{5x+20}{x+4} \times \dfrac{x-3}{5} =$

7) $\dfrac{x-6}{x+7} \times \dfrac{10x+70}{x-6} =$

8) $\dfrac{1}{x+10} \times \dfrac{10x+40}{x+4} =$

9) $\dfrac{5(x+3)}{6x} \times \dfrac{7}{5(x+3)} =$

10) $\dfrac{9(x+2)}{x+2} \times \dfrac{9x}{9(x-7)} =$

11) $\dfrac{3x^2+15x}{x+5} \times \dfrac{1}{x+7} =$

12) $\dfrac{15x^2-15x}{12x^2-12x} \times \dfrac{5x}{5x^2} =$

13) $\dfrac{1}{x-8} \times \dfrac{x^2+5x-24}{x+8} =$

14) $\dfrac{x^2-10x+25}{10x-50} \times \dfrac{x-5}{40-8x} =$

🖎 **Divide.**

15) $\dfrac{8+2x-x^2}{x^2-2x-8} \div \dfrac{2x}{x+6} =$

16) $\dfrac{8a}{a+6} \div \dfrac{8a}{3a+18} =$

17) $\dfrac{12x}{x-5} \div \dfrac{12x}{12x-60} =$

18) $\dfrac{x+10}{7x^2-70x} \div \dfrac{1}{7x} =$

19) $\dfrac{x-23}{x+4x-21} \div \dfrac{13x}{x+8} =$

20) $\dfrac{4x}{x-6} \div \dfrac{4x}{10x-60} =$

21) $\dfrac{x+5}{x+12x+35} \div \dfrac{4x}{x+7} =$

22) $\dfrac{x+3}{x+13x+40} \div \dfrac{6x}{x+8} =$

23) $\dfrac{14x+12}{2} \div \dfrac{63x+54}{2x} =$

24) $\dfrac{8x^3+64x^2}{x^2+13x+40} \div \dfrac{3(x+5)}{3x^3+15x^2} =$

25) $\dfrac{x^2+7x+12}{x^2+5x+6} \div \dfrac{1}{x+2} =$

26) $\dfrac{x^2-x-6}{4x+20} \div \dfrac{2}{x+5} =$

27) $\dfrac{x+4}{x^2+6x+8} \div \dfrac{1}{x+5} =$

28) $\dfrac{1}{3x} \div \dfrac{6x}{3x^2+21x} =$

Algebra 2 Workbook

Solving Rational Equations and Complex Fractions

✎ **Solve each equation. Remember to check for extraneous solutions.**

1) $\dfrac{2x-4}{x+1} = \dfrac{x+8}{x-2}$

2) $\dfrac{1}{x} = \dfrac{3}{4x} + 1$

3) $\dfrac{2x-3}{8x+1} = \dfrac{x+6}{x-2}$

4) $\dfrac{1}{4b^2} + \dfrac{1}{4b} = \dfrac{1}{b^2}$

5) $\dfrac{3x-2}{8x+1} = \dfrac{2x-5}{6x-5}$

6) $\dfrac{1}{n^2} - \dfrac{1}{n} = \dfrac{1}{2n^2}$

7) $\dfrac{1}{6b^2} = \dfrac{1}{3b^2} - \dfrac{1}{b}$

8) $\dfrac{1}{n-6} - 1 = \dfrac{5}{n-6}$

9) $\dfrac{5}{r-1} = -\dfrac{10}{r+1}$

10) $1 = \dfrac{1}{x^2+2x} + \dfrac{x+1}{x}$

11) $\dfrac{1}{x} = 5 + \dfrac{6}{9x}$

12) $\dfrac{x+6}{x^2-3x} - 1 = \dfrac{1}{x^2-3x}$

13) $\dfrac{x-3}{x+4} - 1 = \dfrac{1}{x+3}$

14) $\dfrac{1}{6x^2} = \dfrac{1}{2x^2} - \dfrac{1}{x}$

15) $\dfrac{x+5}{x^2+x} = \dfrac{1}{x^2+x} - \dfrac{x-6}{x+1}$

16) $1 = \dfrac{1}{x^2+2x} + \dfrac{x+1}{x}$

✎ **Simplify each expression.**

17) $\dfrac{\frac{2}{7}}{\frac{2}{35} - \frac{3}{16}} =$

18) $\dfrac{\frac{16}{3}}{-6\frac{2}{13}} =$

19) $\dfrac{8}{\frac{8}{x} + \frac{2}{3x}} =$

20) $\dfrac{x^2}{\frac{4}{6} - \frac{4}{x}} =$

21) $\dfrac{\frac{4}{x-3} - \frac{2}{x+2}}{\frac{6}{x^2+6x+8}} =$

22) $\dfrac{\frac{12}{x-1}}{\frac{12}{6} - \frac{12}{36}} =$

23) $\dfrac{2 + \frac{6}{x-3}}{2 - \frac{3}{x-3}} =$

24) $\dfrac{\frac{1}{2} - \frac{x+3}{4}}{\frac{x^2}{2} - \frac{3}{2}} =$

Answers of Worksheets – Chapter 5

Simplifying and Graphing rational expressions

1) $\frac{1}{2}$

2) $2(x-2)$

3) $\frac{3}{x-1}$

4) $\frac{x+1}{x+4}$

5) $\frac{7x^2}{2}$

6) $\frac{1}{x+4}$

7) $x+2$

8) $\frac{5}{x-1}$

9) Discontinuities: –3, 0;   Vertical Asymptote: $x = -3, x = 2$;   Holes: None
   Horizontal. Asymptote: None;   x–intercepts: 0, –2, 3

10) Discontinuities –1, –3;   Vertical Asymptote $x = -1$;   Holes $x = -3$
    Horizontal Asymptote $y = -\frac{1}{4}$;   x–intercepts. 2

11) Discontinuities: 9;   Vertical Asymptote: $x = 9$;   Holes: None
    Horizontal Asymptote: $y = 1$;   x–intercepts: 2

12) Discontinuities: –2, 3;   Vertical Asymptote: $x = -2, x = 3$;   Holes: None
    Horizontal Asymptote: $y = 0$;   x–intercepts: None

13)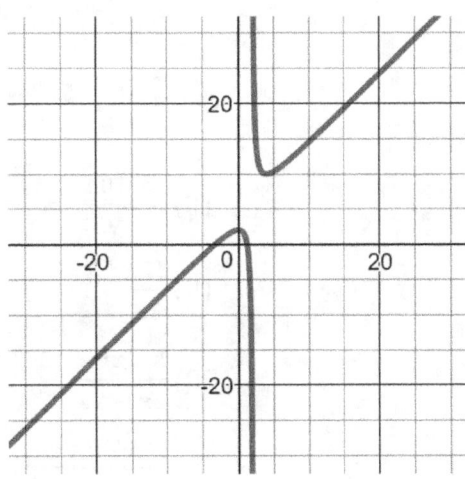

14)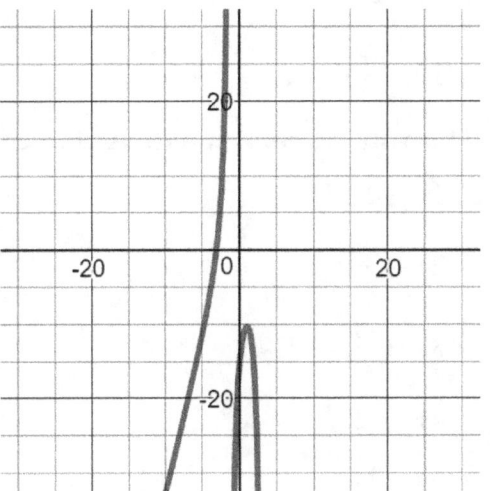

Adding and subtracting rational expressions

1) $\frac{-2+x}{4x+10}$

2) $\frac{2x^2 - x + 21}{(x-4)(x+3)}$

3) $\frac{-x-45}{(x+6)(x-7)}$

4) $\frac{2x-5}{x^2-12}$

5) $\frac{2x+3}{x+3}$

6) $2 + \frac{3}{x}$

7) $\frac{3x}{x+1}$

8) $\frac{24x^2 + 22x}{(5x+4)(2x+3)}$

9) $\frac{-x^2 + 5xy - y^2}{(x-y)(x+y)}$

Algebra 2 Workbook

10) $\dfrac{10}{(x^2-5x+6)(x+3)}$

11) $\dfrac{2x+10}{(x+1)(x+3)}$

12) $\dfrac{2x+2+3x^2}{3\,(3x-4+x^2)}$

13) $\dfrac{4x^2-3x-30}{(x+3)(x-3)}$

14) $\dfrac{4}{(x+1)(x+3)}$

15) $\dfrac{74+12x}{3x\,(x+6)}$

16) $\dfrac{x+1}{2}+2$

17) $\dfrac{44x^2+38x}{(5x+4)(2x+3)}$

18) $\dfrac{8}{5x^2+20x}$

Multiplying and Dividing rational expressions

1) $\dfrac{7}{8}$

2) $\dfrac{494}{35x}$

3) $\dfrac{1144x}{279}$

4) $\dfrac{65x^2}{37}$

5) $\dfrac{62}{21x}$

6) $x-3$

7) 10

8) $\dfrac{10}{x+10}$

9) $\dfrac{7}{6x}$

10) $\dfrac{9x}{x-7}$

11) $\dfrac{3x}{x+7}$

12) $\dfrac{5}{4x}$

13) $\dfrac{x-3}{x-8}$

14) $-\dfrac{(x-5)}{40}$

15) $\dfrac{-x}{2(x+6)}$

16) 3

17) 12

18) $\dfrac{x+10}{x-10}$

19) $\dfrac{x+8}{13x\,(x+7)}$

20) 10

21) $\dfrac{1}{4x}$

22) $\dfrac{x+3}{6x\,(x+5)}$

23) $\dfrac{2x}{9}$

24) $\dfrac{8x^4}{x+5}$

25) $x+4$

26) $\dfrac{(x-3)(x+2)}{8}$

27) $\dfrac{x+5}{x+2}$

28) $\dfrac{x+7}{6x}$

Solving rational equations and complex fractions

1) $\{0,17\}$

2) $\{\tfrac{1}{4}\}$

3) $\{\tfrac{-5}{2},-3\}$

4) $\{3\}$

5) $\{\tfrac{1}{6}\}$

6) $\{\tfrac{1}{2}\}$

7) $\{\tfrac{1}{6}\}$

8) $\{2\}$

9) $\{\tfrac{1}{3}\}$

10) $\{-3\}$

11) $\{\tfrac{1}{15}\}$

12) $\{5,-1\}$

13) $\{-\tfrac{25}{8}\}$

14) $\{\tfrac{1}{3}\}$

15) $\{1,4\}$

16) $\{-3\}$

17) $-\dfrac{160}{73}$

18) $-\dfrac{104}{93}$

19) $\dfrac{12x}{13}$

20) $\dfrac{3x^3}{2x-12}$

21) $\dfrac{(x+7)(x+4)}{3\,(x-3)}$

22) $\dfrac{36}{5x-5}$

23) $\dfrac{2x}{2x-9}$

24) $\dfrac{-1-x}{2x^2-6}$

Chapter 6:
Matrices

Topics that you will practice in this chapter:

- ✓ Adding and Subtracting Matrices
- ✓ Matrix Multiplications
- ✓ Finding Determinants of a Matrix
- ✓ Finding Inverse of a Matrix
- ✓ Matrix Equations

Mathematics is an independent world created out of pure intelligence.

— William Woods Worth

Adding and Subtracting Matrices

✎ **Simplify.**

1) $\begin{vmatrix} -8 & 2 & -2 \end{vmatrix} + \begin{vmatrix} 0 & -2 & -5 \end{vmatrix}$

2) $\begin{vmatrix} 3 & 1 \\ -2 & -1 \\ -5 & 0 \end{vmatrix} + \begin{vmatrix} 2 & -2 \\ 2 & 1 \\ 5 & 3 \end{vmatrix}$

3) $\begin{vmatrix} -1 & 1 & -3 \\ 5 & -3 & -2 \end{vmatrix} - \begin{vmatrix} 7 & -1 & -3 \\ 2 & 3 & -5 \end{vmatrix}$

4) $\begin{vmatrix} 8 & 1 \end{vmatrix} + \begin{vmatrix} -5 & -6 \end{vmatrix}$

5) $\begin{vmatrix} 2 \\ 3 \end{vmatrix} + \begin{vmatrix} 2 \\ 7 \end{vmatrix}$

6) $\begin{vmatrix} -3r + 2t \\ -3r \\ 2s \end{vmatrix} + \begin{vmatrix} 3r \\ -t \\ -5r + 3 \end{vmatrix}$

7) $\begin{vmatrix} 2z - 1 \\ -6 \\ -3 - 6z \\ 5y \end{vmatrix} + \begin{vmatrix} -3y \\ 5z \\ 7 + 2z \\ 6z \end{vmatrix}$

8) $\begin{vmatrix} -3n & n + 7m \\ -n & -5m \end{vmatrix} + \begin{vmatrix} 2 & -3 \\ 3m & 0 \end{vmatrix}$

9) $\begin{vmatrix} 4 & 6 \\ -7 & 2 \end{vmatrix} - \begin{vmatrix} 1 & -2 \\ 3 & 9 \end{vmatrix}$

10) $\begin{vmatrix} 4 & -7 & 6 \\ 3 & -5 & 5 \\ -5 & 5 & -10 \end{vmatrix} + \begin{vmatrix} 0 & 6 & -2 \\ 6 & 4 & 6 \\ 3 & -7 & -4 \end{vmatrix}$

Matrix Multiplication

✎ **Simplify.**

1) $\begin{vmatrix} -2 & -2 \\ -3 & 1 \end{vmatrix} \times \begin{vmatrix} -1 & -2 \\ 2 & 3 \end{vmatrix}$

2) $\begin{vmatrix} 3 & 2 \\ -1 & 0 \\ -2 & 3 \end{vmatrix} \times \begin{vmatrix} -1 & 4 \\ -1 & -3 \end{vmatrix}$

3) $\begin{vmatrix} 3 & 1 & 0 \\ 2 & 5 & 4 \end{vmatrix} \times \begin{vmatrix} 2 & 5 & 1 \\ 1 & -2 & 1 \end{vmatrix}$

4) $\begin{vmatrix} -5 \\ 1 \\ 3 \end{vmatrix} \times \begin{vmatrix} 3 & -4 \end{vmatrix}$

5) $\begin{vmatrix} 5 & -2 \\ 1 & 1 \\ 0 & -4 \end{vmatrix} \times \begin{vmatrix} -2 & 2 \\ 1 & 0 \end{vmatrix}$

6) $\begin{vmatrix} 1 & 1 \\ -3 & 0 \end{vmatrix} \cdot \begin{vmatrix} 5 & -1 \\ 1 & 0 \end{vmatrix}$

7) $\begin{vmatrix} -3 & -2y \\ x & -1 \end{vmatrix} \cdot \begin{vmatrix} -2x & 1 \\ -y & -1 \end{vmatrix}$

8) $\begin{vmatrix} 2 & -3v \end{vmatrix} \cdot \begin{vmatrix} -u & -2v \\ 1 & 0 \end{vmatrix}$

9) $\begin{vmatrix} -2 & 4 & 0 \\ 0 & 2 & -1 \\ 3 & -2 & 3 \\ -3 & 4 & 1 \end{vmatrix} \cdot \begin{vmatrix} 4 & 0 \\ 3 & -3 \\ 2 & 1 \end{vmatrix}$

10) $\begin{vmatrix} 2 & 1 & 0 \\ 0 & 2 & 1 \end{vmatrix} \cdot \begin{vmatrix} -1 & 3 \\ -1 & 4 \\ 3 & -2 \end{vmatrix}$

11) $\begin{vmatrix} -3 & 4 \\ -2 & 3 \end{vmatrix} \cdot \begin{vmatrix} 0 & -2 \\ 3 & 1 \end{vmatrix}$

12) $\begin{vmatrix} 1 & 0 \\ -5 & -2 \end{vmatrix} \cdot \begin{vmatrix} 2 & -2 \\ 1 & 1 \end{vmatrix}$

13) $\begin{vmatrix} 1 & 3 \\ -1 & -2 \end{vmatrix} \cdot \begin{vmatrix} 1 & -4 \\ 0 & 4 \end{vmatrix}$

14) $\begin{vmatrix} -2 & -2 \\ 1 & 0 \\ 2 & 0 \\ 2 & -1 \end{vmatrix} \times \begin{vmatrix} 0 & -1 & 2 \\ -1 & 0 & -4 \end{vmatrix}$

WWW.MathNotion.Com

Finding Determinants of a Matrix

✎ **Evaluate the determinant of each matrix.**

1) $\begin{vmatrix} 4 & 0 \\ -9 & -5 \end{vmatrix}$

2) $\begin{vmatrix} 9 & 5 \\ 1 & 0 \end{vmatrix}$

3) $\begin{vmatrix} -2 & 2 \\ 3 & 3 \end{vmatrix}$

4) $\begin{vmatrix} -1 & 7 \\ -2 & 8 \end{vmatrix}$

5) $\begin{vmatrix} -5 & 2 \\ 2 & 3 \end{vmatrix}$

6) $\begin{vmatrix} 7 & -4 \\ 0 & 8 \end{vmatrix}$

7) $\begin{vmatrix} 1 & -3 \\ 8 & -5 \end{vmatrix}$

8) $\begin{vmatrix} 3 & 5 \\ 4 & 1 \end{vmatrix}$

9) $\begin{vmatrix} 5 & 4 \\ -2 & 8 \end{vmatrix}$

10) $\begin{vmatrix} 3 & 2 \\ 0 & 3 \end{vmatrix}$

11) $\begin{vmatrix} 4 & -1 & 2 \\ 0 & 1 & -2 \\ 2 & 3 & 3 \end{vmatrix}$

12) $\begin{vmatrix} -2 & 1 & -2 \\ -2 & 3 & 1 \\ 1 & 0 & 3 \end{vmatrix}$

13) $\begin{vmatrix} 4 & 2 & 2 \\ 2 & -1 & 3 \\ 0 & 5 & 4 \end{vmatrix}$

14) $\begin{vmatrix} 2 & -1 & 0 \\ 3 & 2 & -2 \\ 2 & 1 & 2 \end{vmatrix}$

15) $\begin{vmatrix} 3 & 1 & 0 \\ 0 & -2 & -2 \\ 1 & 5 & 4 \end{vmatrix}$

Finding Inverse of a Matrix

✎ **Find the inverse of each matrix.**

1) $\begin{vmatrix} 3 & 5 \\ 1 & 4 \end{vmatrix}$

2) $\begin{vmatrix} 3 & 2 \\ 4 & 3 \end{vmatrix}$

3) $\begin{vmatrix} 5 & 4 \\ 2 & 2 \end{vmatrix}$

4) $\begin{vmatrix} 6 & 5 \\ 2 & 4 \end{vmatrix}$

5) $\begin{vmatrix} -4 & 3 \\ 2 & 4 \end{vmatrix}$

6) $\begin{vmatrix} 5 & 2 \\ 7 & 6 \end{vmatrix}$

7) $\begin{vmatrix} 1 & 0 \\ 9 & 4 \end{vmatrix}$

8) $\begin{vmatrix} -8 & -9 \\ 3 & 4 \end{vmatrix}$

9) $\begin{vmatrix} -2 & 7 \\ -2 & 7 \end{vmatrix}$

10) $\begin{vmatrix} -3 & 2 \\ 5 & 4 \end{vmatrix}$

11) $\begin{vmatrix} 8 & 4 \\ 1 & 2 \end{vmatrix}$

12) $\begin{vmatrix} 1 & 8 \\ 2 & 0 \end{vmatrix}$

13) $\begin{vmatrix} 1 & 9 \\ 0 & 0 \end{vmatrix}$

14) $\begin{vmatrix} 10 & 6 \\ 5 & 3 \end{vmatrix}$

Matrix Equations

✎ **Solve each equation.**

1) $\begin{vmatrix} -2 & 4 \\ 0 & 1 \end{vmatrix} z = \begin{vmatrix} 8 \\ 5 \end{vmatrix}$

2) $3x = \begin{vmatrix} 15 & -6 \\ 9 & -12 \end{vmatrix}$

3) $\begin{vmatrix} -4 & 3 \\ -9 & 5 \end{vmatrix} = \begin{vmatrix} 1 & 6 \\ 3 & 7 \end{vmatrix} - x$

4) $Y - \begin{vmatrix} -3 \\ -5 \\ 11 \\ 11 \end{vmatrix} = \begin{vmatrix} -2 \\ 8 \\ -18 \\ -2 \end{vmatrix}$

5) $\begin{vmatrix} -2 & -1 \\ 1 & -3 \end{vmatrix} C = \begin{vmatrix} 5 \\ -6 \end{vmatrix}$

6) $\begin{vmatrix} -1 & -2 \\ 4 & 3 \end{vmatrix} B = \begin{vmatrix} 0 & -1 & -1 \\ -5 & 14 & -1 \end{vmatrix}$

7) $\begin{vmatrix} -2 & 2 \\ 3 & -1 \end{vmatrix} C = \begin{vmatrix} 10 \\ -9 \end{vmatrix}$

8) $\begin{vmatrix} 2 & 5 \\ 1 & 1 \end{vmatrix} C = \begin{vmatrix} 4 \\ -1 \end{vmatrix}$

9) $\begin{vmatrix} 0 & -5 \\ 2 & 4 \end{vmatrix} Z = \begin{vmatrix} 15 \\ 0 \end{vmatrix}$

10) $\begin{vmatrix} -9 \\ 6 \\ -15 \end{vmatrix} = 3B$

11) $\begin{vmatrix} -7 \\ 4 \\ 2 \end{vmatrix} = y - \begin{vmatrix} 5 \\ -4 \\ -8 \end{vmatrix}$

12) $-3b - \begin{vmatrix} 8 \\ 4 \\ -2 \end{vmatrix} = \begin{vmatrix} -26 \\ -7 \\ -16 \end{vmatrix}$

Answers of Worksheets – Chapter 6

Adding and Subtracting Matrices

1) $\begin{vmatrix} -8 & 0 & -7 \end{vmatrix}$

2) $\begin{vmatrix} 5 & -1 \\ 0 & 0 \\ 0 & 3 \end{vmatrix}$

3) $\begin{vmatrix} -8 & 2 & 0 \\ 3 & -6 & 3 \end{vmatrix}$

4) $\begin{vmatrix} 3 & -5 \end{vmatrix}$

5) $\begin{vmatrix} 4 \\ 10 \end{vmatrix}$

6) $\begin{vmatrix} 2t \\ -3r - t \\ 2s - 5r + 3 \end{vmatrix}$

7) $\begin{vmatrix} 2z - 1 - 3y \\ -6 + 5z \\ 4 - 4z \\ 5y + 6z \end{vmatrix}$

8) $\begin{vmatrix} -3n + 2 & n + 7m - 3 \\ -n + 3m & -5m \end{vmatrix}$

9) $\begin{vmatrix} 3 & 8 \\ -10 & -7 \end{vmatrix}$

10) $\begin{vmatrix} 4 & -1 & 4 \\ 9 & -1 & 11 \\ -2 & -2 & -14 \end{vmatrix}$

Matrix Multiplication

1) $\begin{vmatrix} -2 & -2 \\ 5 & 9 \end{vmatrix}$

2) $\begin{vmatrix} -5 & 6 \\ 1 & -4 \\ -1 & -17 \end{vmatrix}$

3) Undefined

4) $\begin{vmatrix} -15 & 20 \\ 3 & -4 \\ 9 & -12 \end{vmatrix}$

5) $\begin{vmatrix} -12 & 10 \\ -1 & 2 \\ -4 & 0 \end{vmatrix}$

6) $\begin{vmatrix} 6 & -1 \\ -15 & 3 \end{vmatrix}$

7) $\begin{vmatrix} 6x + 2y^2 & 2y - 3 \\ -2x^2 + y & x + 1 \end{vmatrix}$

8) $\begin{vmatrix} -2u - 3v & -4v \end{vmatrix}$

9) $\begin{vmatrix} 4 & -12 \\ 4 & -7 \\ 12 & 9 \\ 2 & -11 \end{vmatrix}$

10) $\begin{vmatrix} -3 & 10 \\ 1 & 6 \end{vmatrix}$

11) $\begin{vmatrix} 12 & 10 \\ 9 & 7 \end{vmatrix}$

12) $\begin{vmatrix} 2 & -2 \\ -12 & 8 \end{vmatrix}$

13) $\begin{vmatrix} 1 & 8 \\ -1 & -4 \end{vmatrix}$

14) $\begin{vmatrix} 2 & 2 & 4 \\ 0 & -1 & 2 \\ 0 & -2 & 4 \\ 1 & -2 & 8 \end{vmatrix}$

Finding Determinants of a Matrix

1) –20
2) –5
3) –12
4) 6
5) –19
6) 56
7) 19
8) –17
9) 48
10) 9
11) 36
12) –5

13) −72

14) 22

15) 4

Finding Inverse of a Matrix

1) $\begin{vmatrix} \frac{4}{7} & \frac{-5}{7} \\ \frac{-1}{7} & \frac{3}{7} \end{vmatrix}$

2) $\begin{vmatrix} 3 & -2 \\ -4 & 3 \end{vmatrix}$

3) $\begin{vmatrix} 1 & -2 \\ -1 & \frac{5}{2} \end{vmatrix}$

4) $\begin{vmatrix} \frac{2}{7} & \frac{-5}{14} \\ \frac{-1}{7} & \frac{3}{7} \end{vmatrix}$

5) $\begin{vmatrix} -\frac{2}{11} & \frac{3}{22} \\ \frac{1}{11} & \frac{2}{11} \end{vmatrix}$

6) $\begin{vmatrix} \frac{3}{8} & -\frac{1}{8} \\ -\frac{5}{16} & \frac{5}{16} \end{vmatrix}$

7) $\begin{vmatrix} 1 & 0 \\ -\frac{9}{4} & \frac{1}{4} \end{vmatrix}$

8) $\begin{vmatrix} -\frac{4}{5} & -\frac{9}{5} \\ \frac{3}{5} & \frac{8}{5} \end{vmatrix}$

9) No inverse exists.

10) $\begin{vmatrix} -\frac{2}{11} & \frac{1}{11} \\ \frac{5}{22} & \frac{3}{22} \end{vmatrix}$

11) $\begin{vmatrix} \frac{1}{6} & -\frac{1}{3} \\ -\frac{1}{12} & \frac{2}{3} \end{vmatrix}$

12) $\begin{vmatrix} 0 & \frac{1}{2} \\ \frac{1}{8} & -\frac{1}{16} \end{vmatrix}$

13) No inverse exists.

14) No inverse exists.

Matrix Equations

1) $\begin{vmatrix} 6 \\ 5 \end{vmatrix}$

2) $\begin{vmatrix} 5 & -2 \\ 3 & -4 \end{vmatrix}$

3) $\begin{vmatrix} 5 & 3 \\ 12 & 2 \end{vmatrix}$

4) $\begin{vmatrix} -5 \\ 3 \\ -7 \\ 9 \end{vmatrix}$

5) $\begin{vmatrix} -3 \\ 1 \end{vmatrix}$

6) $\begin{vmatrix} -2 & 5 & -1 \\ 1 & -2 & 1 \end{vmatrix}$

7) $\begin{vmatrix} -2 \\ 3 \end{vmatrix}$

8) $\begin{vmatrix} -3 \\ 2 \end{vmatrix}$

9) $\begin{vmatrix} 6 \\ -3 \end{vmatrix}$

10) $\begin{vmatrix} -3 \\ 2 \\ -5 \end{vmatrix}$

11) $\begin{vmatrix} -2 \\ 0 \\ -6 \end{vmatrix}$

12) $\begin{vmatrix} 6 \\ 1 \\ 6 \end{vmatrix}$

Chapter 7:
Sequences and Series

Topics that you will practice in this chapter:
- ✓ Arithmetic Sequences
- ✓ Geometric Sequences
- ✓ Comparing Arithmetic and Geometric Sequences
- ✓ Finite Geometric Series
- ✓ Infinite Geometric Series

Mathematics is like checkers in being suitable for the young, not too difficult, amusing, and without peril to the state. — Plato

Arithmetic Sequences

✎ **Find the next three terms of each arithmetic sequence.**

1) 32, 23, 14, 5, −4, ...

2) −91, −63, −35, −7, ...

3) 51, 62, 73, 84, 95, ...

4) 84, 53, 22, −9, −40, ...

✎ **Given the first term and the common difference of an arithmetic sequence find the first five terms and the explicit formula.**

5) $a_1 = 9, d = 12$

6) $a_1 = -10, d = -5$

7) $a_1 = 52, d = 22$

8) $a_1 = 210, d = -102$

✎ **Given a term in an arithmetic sequence and the common difference find the first five terms and the explicit formula.**

9) $a_{51} = -468, d = -12$

10) $a_{31} = 230, d = 6$

11) $a_{62} = -128.2, d = -4.2$

12) $a_{33} = -2,352, d = -77$

✎ **Given a term in an arithmetic sequence and the common difference find the recursive formula and the three terms in the sequence after the last one given.**

13) $a_{25} = -156, d = -6$

14) $a_{16} = 111, d = 7.1$

15) $a_{22} = 43, d = 1.8$

16) $a_{14} = -17, d = 0.4$

Geometric Sequences

✎ **Determine if the sequence is geometric. If it is, find the common ratio.**

1) $2, -14, 98, -686, \ldots$

2) $-3, -15, -75, -375, \ldots$

3) $7, 17, 31, 126, \ldots$

4) $-5, -35, -245, -1715, \ldots$

✎ **Given the first term and the common ratio of a geometric sequence find the first five terms and the explicit formula.**

5) $a_1 = 0.7, r = -3$

6) $a_1 = 0.4, r = 5$

✎ **Given the recursive formula for a geometric sequence find the common ratio, the first five terms, and the explicit formula.**

7) $a_n = a_{n-1} \times 6, a_1 = 2$

8) $a_n = a_{n-1} \cdot (-4), a_1 = -6$

9) $a_n = a_{n-1} \cdot 9, a_1 = 0.2$

10) $a_n = a_{n-1} \cdot 3, a_1 = -8$

✎ **Given two terms in a geometric sequence find the 9th term and the recursive formula.**

11) $a_5 = 729$ and $a_6 = -243$

12) $a_6 = -768$ and $a_3 = 12$

Comparing Arithmetic and Geometric Sequences

✎ **For each sequence, state if it is arithmetic, geometric, or neither.**

1) $5, 10, 15, 20, \ldots$

2) $6, 10, 14, 20, \ldots$

3) $2, 6, 24, 51, \ldots$

4) $1, 8, 18, 28, 36, \ldots$

5) $2, 8, 17, 52, 142, \ldots$

6) $2, 5, 9, 17, 36, \ldots$

7) $0.6, 3, 15, 75, 375, \ldots$

8) $4, 20, 100, 500, \ldots$

9) $-18, -23, -28, -33, -38, \ldots$

10) $-3, 12, -48, 192, -768, \ldots$

11) $8, 18, 26, 39, 50, \ldots$

12) $3, 12, 90, 150, 210 \ldots$

13) $-22, -12, -2, 2, 12, \ldots$

14) $a_n = 2 \cdot 7^{n-1}$

15) $a_n = 8 \cdot 4^{n-1}$

16) $a_n = 9 - 5n$

17) $a_n = -110 + 200n$

18) $a_n = 15 + 13n$

19) $a_n = -6 \cdot (-11)^{n-1}$

20) $a_n = 120 + 42n$

21) $a_n = (4n)^4$

22) $a_n = 28 + 6n$

23) $a_n = -(13)^{n-1}$

24) $a_n = -7 \cdot (1.5)^{n-1}$

25) $a_n = \frac{2n+1}{7n}$

26) $a_n = \frac{24-13n}{6n}$

27) $a_n = \frac{8-17n}{2n}$

28) $a_n = \frac{32-a_{n-1}}{9}$

29) $a_n = -\frac{3}{19} + \frac{2}{7}n$

Finite Geometric Series

✍ **Evaluate the related series of each sequence.**

1) $-2, 8, -32, 128$

2) $-1, 3, -9, 27, -81$

3) $-1, 4, -16, 64, -256$

4) $1, 8, 64, 512$

5) $-6, -24, -96, -384$

6) $2, -12, 72, -432, 2592$

✍ **Evaluate each geometric series described.**

7) $1 + 3 + 9 + 27 \ldots, n = 6$ _____

8) $1.5 - 6 + 24 - 96 \ldots, n = 6$ _____

9) $-2 - 6 - 18 - 54 \ldots, n = 7$ _____

10) $0.5 - 3 + 18 - 108 \ldots, n = 6$ _____

11) $2.5 - 10 + 40 - 160 \ldots, n = 8$ _____

12) $-1 + 7 - 49 + 343 \ldots, n = 6$ _____

13) $a_1 = -2, r = 6, n = 5$ _____

14) $a_1 = 3, r = 2, n = 9$ _____

15) $\sum_{n=1}^{5} 4 \cdot (-3)^{n-1}$ _____

16) $\sum_{n=1}^{7} 6 \cdot (-2)^{n-1}$ _____

17) $\sum_{n=1}^{5} 3 \cdot (5)^{n-1}$ _____

18) $\sum_{m=1}^{10} (-2)^{m-1}$ _____

19) $\sum_{m=1}^{4} 8 \times (5)^{m-1}$ _____

20) $\sum_{k=1}^{8} 2 \times (4)^{k-1}$ _____

Infinite Geometric Series

✍ **Determine if each geometric series converges or diverges.**

1) $a_1 = -1.4, \ r = 6$

2) $a_1 = 5.2, r = 0.3$

3) $a_1 = -6, r = 7.2$

4) $a_1 = 12, r = 0.04$

5) $a_1 = 3, r = 15$

6) $-1, 7, -49, 343, \ldots$

7) $6, -1, \frac{1}{6}, -\frac{1}{36}, \frac{1}{216}, \ldots$

8) $512 + 64 + 8 + 1 \ldots$

9) $-4 + \frac{12}{7} - \frac{36}{49} + \frac{108}{343} \ldots$

10) $\frac{120}{459} - \frac{60}{153} + \frac{30}{51} - \frac{15}{17} \ldots$

✍ **Evaluate each infinite geometric series described.**

11) $a_1 = 4, r = -\frac{1}{6}$

12) $a_1 = 18, r = -\frac{1}{3}$

13) $a_1 = 16, r = -\frac{1}{7}$

14) $a_1 = 8, r = \frac{1}{3}$

15) $2 + 0.5 + 0.125 + 0.031 + \cdots$

16) $125 - 25 + 5 - 1 \ldots,$

17) $1 - 0.6 + 0.36 - 0.216 \ldots,$

18) $3 + \frac{12}{5} + \frac{48}{25} + \frac{192}{125} \ldots,$

19) $\sum_{k=1}^{\infty} 11^{k-1}$

20) $\sum_{i=1}^{\infty} \left(\frac{2}{5}\right)^{i-1}$

21) $\sum_{k=1}^{\infty} \left(-\frac{3}{7}\right)^{k-1}$

22) $\sum_{n=1}^{\infty} 12\left(\frac{5}{6}\right)^{n-1}$

Answers of Worksheets – Chapter 7

Arithmetic Sequences

1) $-13, -22, -31$

2) $21, 49, 77$

3) $106, 117, 128$

4) $-71, -102, -133$

5) First Five Terms: $9, 21, 33, 45, 57$, Explicit: $a_n = 9 + 12(n-1)$

6) First Five Terms: $-10, -15, -20, -25, -30$, Explicit: $a_n = -10 - 5(n-1)$

7) First Five Terms: $52, 74, 96, 118, 140$, Explicit: $a_n = 52 + 22(n-1)$

8) First Five Terms: $210, 108, 6, -96, -198$, Explicit: $a_n = 210 - 102(n-1)$

9) First Five Terms: $132, 120, 108, 96, 84$, Explicit: $a_n = 132 - 12(n-1)$

10) First Five Terms: $50, 56, 62, 68, 74$, Explicit: $a_n = 50 + 6(n-1)$

11) First Five Terms: $128, 123.8, 119.6, 115.4, 111.2$, Explicit: $a_n = 128 - 4.2(n-1)$

12) First Five Terms: $112, 35, -42, -119, -196$, Explicit: $a_n = 112 - 77(n-1)$

13) Next 3 terms: $-162, -168, -174$, Recursive: $a_n = a_{n-1} - 6, a_1 = -6$

14) Next 3 terms: $118.2, 125.2, 132.3$ Recursive: $a_n = a_{n-1} + 7.1, a_1 = 4.5$

15) Next 3 terms: $44.8, 46.6, 48.4$, Recursive: $a_n = a_{n-1} + 1.8, a_1 = 5.2$

16) Next 3 terms: $-9.8, -9.6, -9.4$, Recursive: $a_n = a_{n-1} + 0.4, a_1 = -22.2$

Geometric Sequences

1) $r = -7$

2) $r = 5$

3) not geometric

4) $r = 7$

5) First Five Terms: $0.7, -2.1, 6.3, -18.9, 56.7$

   Explicit: $a_n = 0.7 \times (-3)^{n-1}$

6) First Five Terms: $0.4, 2, 10, 50, 250$

   Explicit: $a_n = 0.4 \times (5)^{n-1}$

7) Common Ratio: $r = 6$

   First Five Terms: $2, 12, 72, 432, 2{,}592$

   Explicit: $a_n = 2 \cdot (6)^{n-1}$

8) Common Ratio: $r = -4$
   First Five Terms: $-6, 24, -96, 384, -1,536$
   Explicit: $a_n = -6 \cdot (-4)^{n-1}$

9) Common Ratio: $r = 9$
   First Five Terms: $0.2;\ 1.8;\ 16.2;\ 145.8;\ 1,312.2;\ 11,809.8$
   Explicit: $a_n = 0.2 \cdot (9)^{n-1}$

10) Common Ratio: $r = 3$
    First Five Terms: $-8, -24, -72, -216, -648$
    Explicit: $a_n = -8 \cdot (3)^{n-1}$

11) $a_9 = 9$, Recursive: $a_n = a_{n-1} \cdot (\frac{-1}{3})$, $a_1 = 59,049$

12) $a_9 = 49,152$, Recursive: $a_n = a_{n-1} \cdot (-4)$, $a_1 = 0.75$

Comparing Arithmetic and Geometric Sequences

1) Arithmetic	11) Neither	21) Neither
2) Arithmetic	12) Neither	22) Arithmetic
3) Neither	13) Arithmetic	23) Geometric
4) Neither	14) Geometric	24) Geometric
5) Neither	15) Geometric	25) Neither
6) Neither	16) Arithmetic	26) Neither
7) Geometric	17) Arithmetic	27) Neither
8) Geometric	18) Arithmetic	28) Neither
9) Arithmetic	19) Geometric	29) Arithmetic
10) Geometric	20) Arithmetic	

Finite Geometric

1) 102	8) $-1,228.5$	15) 244
2) -61	9) $-2,186$	16) 258
3) -205	10) $-3,333.5$	17) 2,343
4) 585	11) $-32,767.5$	18) -341
5) -510	12) 14,699	19) 1,248
6) 2,157	13) $-3,110$	20) 43,680
7) 364	14) 1,533	

Infinite Geometric

1) Diverges
2) Converges
3) Diverges
4) Converges
5) Diverges
6) Diverges
7) Converges
8) Converges
9) Converges
10) Diverges
11) $\frac{24}{7}$
12) $\frac{27}{2}$
13) 14
14) 12
15) $\frac{8}{3}$
16) $\frac{625}{6}$
17) $\frac{5}{8}$
18) 15
19) Infinite
20) $\frac{5}{3}$
21) $\frac{7}{10}$
22) 72

Chapter 8:
Complex Numbers

Topics that you will practice in this chapter:

- ✓ Adding and Subtracting Complex Numbers
- ✓ Multiplying and Dividing Complex Numbers
- ✓ Graphing Complex Numbers
- ✓ Rationalizing Imaginary Denominators

Mathematics is a hard thing to love. It has the unfortunate habit, like a rude dog, of turning its most unfavorable side towards you when you first make contact with it. — David Whiteland

Adding and Subtracting Complex Numbers

✏️ **Simplify.**

1) $(8i) - (4i) =$

2) $(5i) + (2i) =$

3) $(2i) + (7i) =$

4) $(-6i) - (i) =$

5) $(12i) + (4i) =$

6) $(4i) - (-4i) =$

7) $(-4i) + (-5i) =$

8) $(13i) - (6i) =$

9) $(-21i) - (7i) =$

10) $(-4i) + (2 + 8i) =$

11) $(8 - 4i) + (-6i) =$

12) $(-3i) + (9 + 12i) =$

13) $5 + (9 - 2i) =$

14) $(10i) - (-6 + 2i) =$

15) $(3 + 9i) - (-4i) =$

16) $(7 + 8i) + (-5i) =$

17) $(5i) - (-3i + 4) =$

18) $(6i + 2) + (-2i) =$

19) $(12) - (18 + 3i) =$

20) $(7 + 3i) + (6 + 2i) =$

21) $(4 - 9i) + (3 + 8i) =$

22) $(7 + 3i) + (10 + 12i) =$

23) $(-5 + 5i) - (-5 - 7i) =$

24) $(-8 + 12i) - (-9 + 8i) =$

25) $(-18 + 3i) - (-3 - 12i) =$

26) $(-13 - 4i) + (9 + 12i) =$

27) $(-15 - 2i) - (-14 - 6i) =$

28) $-4 + (8i) + (-14 + 7i) =$

29) $19 - (8i) + (2 - 5i) =$

30) $-3 + (-4 - 8i) - 9 =$

31) $(-24i) + (5 - 8i) + 12 =$

32) $(-11i) - (15 - 12i) + 9i =$

Multiplying and Dividing Complex Numbers

✎ **Simplify.**

1) $(5i)(-3i) =$

2) $(-7i)(2i) =$

3) $(3i)(3i)(-3i) =$

4) $(6i)(-6i) =$

5) $(-7-6i)(7+6i) =$

6) $(4-2i)^2 =$

7) $(5-2i)(4-2i) =$

8) $(5+2i)^2 =$

9) $(7i)(-3i)(9-2i) =$

10) $(2-8i)(6-8i) =$

11) $(-9+3i)(1+4i) =$

12) $(7-8i)(9-3i) =$

13) $5(3i) - (5i)(-4+2i) =$

14) $\dfrac{5}{-25i} =$

15) $\dfrac{12-9i}{-3i} =$

16) $\dfrac{4+9i}{i} =$

17) $\dfrac{20i}{-6+2i} =$

18) $\dfrac{-4-11i}{2i} =$

19) $\dfrac{7i}{3-i} =$

20) $\dfrac{4-9i}{12-5i} =$

21) $\dfrac{8-3i}{-4-4i} =$

22) $\dfrac{-9-5i}{-3-i} =$

23) $\dfrac{-4+i}{-6-5i} =$

24) $\dfrac{-6-7i}{-3+4i} =$

25) $\dfrac{8+11i}{5-5i} =$

Algebra 2 Workbook

Graphing Complex Numbers

✍ **Identify each complex number graphed.**

1)

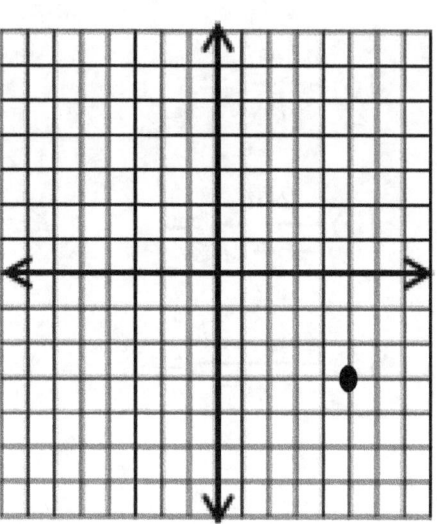

2)

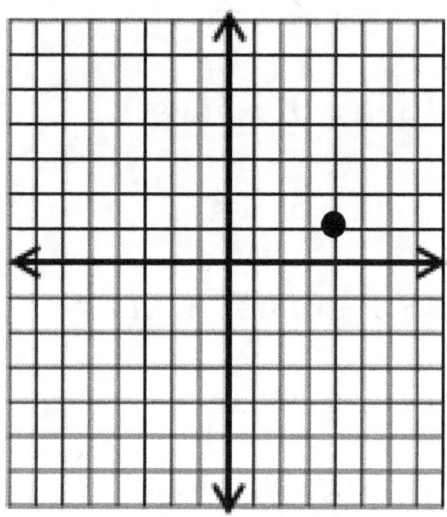

3)

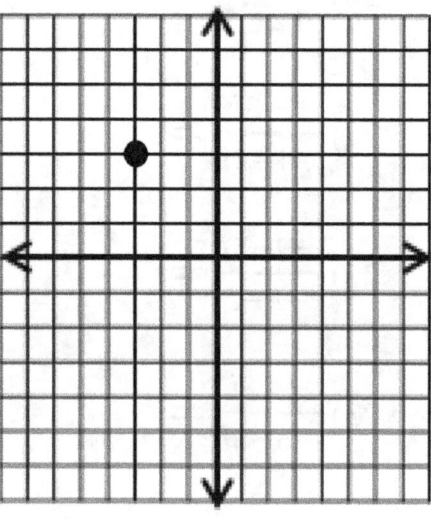

4)

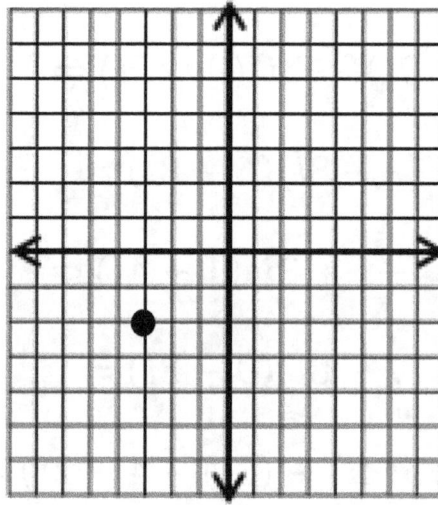

WWW.MathNotion.Com

Rationalizing Imaginary Denominators

✎ **Simplify.**

1) $\dfrac{-8}{-8i} =$

2) $\dfrac{-3}{-21i} =$

3) $\dfrac{-14}{-28i} =$

4) $\dfrac{-30}{-5i} =$

5) $\dfrac{9}{2i} =$

6) $\dfrac{27}{-9i} =$

7) $\dfrac{45}{-20i} =$

8) $\dfrac{-26}{8i} =$

9) $\dfrac{6x}{3yi} =$

10) $\dfrac{9-9i}{-3i} =$

11) $\dfrac{4-9i}{-i} =$

12) $\dfrac{12+4i}{3i} =$

13) $\dfrac{8i}{-1+4i} =$

14) $\dfrac{8i}{-2+6i} =$

15) $\dfrac{-15-3i}{-4+4i} =$

16) $\dfrac{-5-9i}{3+4i} =$

17) $\dfrac{-11-4i}{5-2i} =$

18) $\dfrac{-4+6i}{-3i} =$

19) $\dfrac{9+5i}{4i} =$

20) $\dfrac{-5-3i}{7-2i} =$

21) $\dfrac{-8+i}{-3i} =$

22) $\dfrac{9+i}{-5-2i} =$

23) $\dfrac{-9-5i}{-7-2i} =$

24) $\dfrac{9i-5}{-3-6i} =$

Answers of Worksheets – Chapter 8

Adding and Subtracting Complex Numbers

1) $4i$
2) $7i$
3) $9i$
4) $-7i$
5) $16i$
6) $8i$
7) $-9i$
8) $7i$
9) $-28i$
10) $2 + 4i$
11) $8 - 10i$
12) $9 + 9i$
13) $14 - 2i$
14) $6 + 8i$
15) $3 + 13i$
16) $7 + 3i$
17) $-4 + 8i$
18) $2 + 4i$
19) $-6 - 3i$
20) $13 + 5i$
21) $7 - i$
22) $17 + 15i$
23) $12i$
24) $1 + 4i$
25) $-15 + 15i$
26) $-4 + 8i$
27) $-1 + 4i$
28) $-18 + 15i$
29) $21 - 13i$
30) $-16 - 8i$
31) $17 - 32i$
32) $-15 + 10i$

Multiplying and Dividing Complex Numbers

1) 15
2) 14
3) $27i$
4) 36
5) $-13 - 84i$
6) $-16i + 12$
7) $16 - 18i$
8) $21 + 20i$
9) $189 - 42i$
10) $-52 - 64i$
11) $-21 - 33i$
12) $39 - 93i$
13) $10 + 35i$
14) $\frac{i}{5}$
15) $3 + 4i$
16) $9 - 4i$
17) $1 - 3i$
18) $\frac{11}{2} - 2i$
19) $-\frac{7}{10} + \frac{21}{10}i$
20) $\frac{93}{169} - \frac{88}{169}i$
21) $-\frac{5}{8} + \frac{11}{8}i$
22) $\frac{16}{5} + \frac{3}{5}i$
23) $\frac{19}{61} - \frac{26}{61}i$
24) $-\frac{2}{5} + \frac{9}{5}i$
25) $-\frac{3}{10} + \frac{19}{10}i$

Graphing Complex Numbers

2) $5 - 3i$
3) $4 + i$
4) $-3 + 3i$
5) $-3 - 2i$

Rationalizing Imaginary Denominators

1) $-i$
2) $-\frac{1}{7}i$
3) $\frac{-1}{2}i$
4) $-6i$
5) $-\frac{9}{2}i$
6) $3i$
7) $\frac{9}{4}i$
8) $\frac{13}{4}i$
9) $-\frac{2x}{y}i$
10) $3 + 3i$
11) $9 + 4i$
12) $-\frac{4}{3} + 4i$
13) $\frac{32}{17} - \frac{8}{17}i$
14) $\frac{6}{5} - \frac{2}{5}i$
15) $\frac{3}{2} + \frac{9}{4}i$
16) $-\frac{51}{25} - \frac{7}{25}i$
17) $-\frac{47}{29} - \frac{42}{29}i$
18) $-2 - \frac{4}{3}i$
19) $-\frac{5}{4} + \frac{9}{4}i$
20) $-\frac{29}{53} - \frac{31}{53}i$
21) $-\frac{1}{3} - \frac{8}{3}i$
22) $-\frac{47}{29} + \frac{13}{29}i$
23) $\frac{73}{53} + \frac{17}{53}i$
24) $-\frac{13}{15} - \frac{19}{15}i$

Chapter 9: Logarithms

Topics that you will practice in this chapter:

- ✓ Rewriting Logarithms
- ✓ Evaluating Logarithms
- ✓ Properties of Logarithms
- ✓ Natural Logarithms
- ✓ Exponential Equations Requiring Logarithms
- ✓ Solving Logarithmic Equations

Mathematics is an art of human understanding. — William Thurston

Rewriting Logarithms

✎ **Rewrite each equation in exponential form.**

1) $\log_5 25 = 2$

2) $\log_4 256 = 4$

3) $\log_3 81 = 4$

4) $\log_8 64 = 2$

5) $\log_6 216 = 3$

6) $\log_2 16 = 4$

7) $\log_{10} 100 = 2$

8) $\log_3 243 = 5$

9) $\log_5 625 = 4$

10) $\log_2 256 = 8$

11) $\log_3 6{,}561 = 8$

12) $\log_{11} 121 = 2$

13) $\log_{14} 196 = 2$

14) $\log_{81} 3 = \frac{1}{4}$

15) $\log_{27} 3 = \frac{1}{3}$

16) $\log_{32} 2 = \frac{1}{5}$

17) $\log_{512} 8 = \frac{1}{3}$

18) $\log_2 \frac{1}{8} = -3$

19) $\log_2 \frac{1}{16} = -4$

20) $\log_a \frac{7}{3} = b$

✎ **Rewrite each exponential equation in logarithmic form.**

21) $12^2 = 144$

22) $7^3 = 343$

23) $4^5 = 1{,}024$

24) $15^2 = 225$

25) $5^4 = 625$

26) $6^4 = 1{,}296$

27) $2^9 = 512$

28) $5^5 = 3{,}125$

29) $4^{-6} = \frac{1}{4{,}096}$

30) $3^{-5} = \frac{1}{243}$

31) $16^{-2} = \frac{1}{256}$

32) $6^{-3} = \frac{1}{216}$

33) $3^{-9} = \frac{1}{19{,}683}$

34) $21^{-2} = \frac{1}{441}$

Algebra 2 Workbook

Evaluating Logarithms

✎ **Evaluate each logarithm.**

1) $\log_3 2{,}187 =$

2) $\log_2 256 =$

3) $\log_5 125 =$

4) $\log_5 625 =$

5) $\log_3 243 =$

6) $\log_4 1{,}024 =$

7) $\log_8 64 =$

8) $\log_8 \frac{1}{8} =$

9) $\log_6 \frac{1}{36} =$

10) $\log_2 \frac{1}{16} =$

11) $\log_6 \frac{1}{216} =$

12) $\log_3 \frac{1}{256} =$

13) $\log_{18} \frac{1}{324} =$

14) $\log_{256} \frac{1}{4} =$

15) $\log_{512} 8 =$

16) $\log_4 \frac{1}{4{,}096} =$

17) $\log_9 \frac{1}{729} =$

18) $\log_{216} \frac{1}{6} =$

✎ **Circle the points which are on the graph of the given logarithmic functions.**

19) $y = 5\log_8(3x - 4) + 1$ $(6, 5)$, $(4, 6)$, $(4, 8)$

20) $y = 3\log_2(4x) - 6$ $(4, 6)$, $(\frac{1}{4}, 16)$, $(\frac{1}{4}, -6)$

21) $y = -3\log_5(x - 2) + 5$ $(7, -2)$, $(7, 2)$, $(6, -3)$

22) $y = \frac{1}{2}\log_6(6x) + 4$ $(6, 5)$, $(6, \frac{1}{5})$, $(6, -5)$

23) $y = -\log_9 9(x + 5) + 4$ $(-4, 2)$, $(4, 0)$, $(4, 2)$

24) $y = -\log_8(x - 6) - 4$ $(7, -\frac{1}{4})$, $(7, -4)$, $(7, -\frac{1}{4})$

25) $y = -3\log_6(x + 3) + 6$ $(3, 3)$, $(-5, -3)$, $(-5, 3)$

WWW.MathNotion.Com

Properties of Logarithms

✍ **Expand each logarithm.**

1) $\log(9 \times 4) =$

2) $\log(6 \times 3) =$

3) $\log(2 \times 8) =$

4) $\log\left(\frac{8}{7}\right) =$

5) $\log\left(\frac{9}{5}\right) =$

6) $\log\left(\frac{4}{11}\right)^3 =$

7) $\log(9 \times 4^3) =$

8) $\log\left(\frac{7}{3}\right)^2 =$

9) $\log\left(\frac{5^4}{9}\right) =$

10) $\log(x \times y)^7 =$

11) $\log(x^2 \times y \times z^5) =$

12) $\log\left(\frac{u^8}{v}\right) =$

13) $\log\left(\frac{x}{y^4}\right) =$

✍ **Condense each expression to a single logarithm.**

14) $\log 7 - \log 12 =$

15) $\log 8 + \log 3 =$

16) $4 \log 2 - 7 \log 5 =$

17) $6 \log 4 - 9 \log 5 =$

18) $3 \log 8 - \log 17 =$

19) $8 \log 3 - 6 \log 2 =$

20) $\log 11 - 2 \log 5 =$

21) $4 \log 6 + 3 \log 9 =$

22) $4 \log 5 + 5 \log 13 =$

23) $7 \log_5 a + 16 \log_5 b =$

24) $5 \log_6 x - 7 \log_6 y =$

25) $\log_5 u - 9 \log_5 v =$

26) $8 \log_3 u + 21 \log_3 v =$

27) $26 \log_7 u - 15 \log_7 v =$

Natural Logarithms

✍ **Solve each equation for** x.

1) $e^x = 9$
2) $e^x = 36$
3) $e^x = 49$
4) $\ln x = 3$
5) $\ln(\ln x) = 3$
6) $e^x = 11$
7) $\ln(5x + 9) = 1$
8) $\ln(7x + 3) = 3$
9) $\ln(8x + 5) = 4$
10) $\ln x = \frac{1}{3}$
11) $\ln 6x = e^4$
12) $\ln x = \ln 4 + \ln 7$
13) $\ln x = 3\ln 3 + \ln 8$

✍ **Evaluate without using a calculator.**

14) $3\ln e =$
15) $\ln e^{10} =$
16) $4 \ln e =$
17) $\ln e^{15} =$
18) $13\ln e =$
19) $3\ln e^4 =$
20) $e^{\ln 19} =$
21) $e^{2\ln 5} =$
22) $e^{4\ln 3} =$
23) $\ln \sqrt[6]{e} =$

✍ **Reduce the following expressions to simplest form.**

24) $e^{-2\ln 6 + 2\ln 4} =$
25) $e^{-2\ln\left(\frac{4}{5e}\right)} =$
26) $2\ln(e^6) =$
27) $\ln\left(\frac{1}{e}\right)^9 =$
28) $e^{\ln 6 + 3\ln 5} =$
29) $e^{\ln\left(\frac{13}{e}\right)} =$
30) $7\ln(1^{-3e}) =$
31) $\ln\left(\frac{1}{e}\right)^{-12} =$
32) $3\ln\left(\frac{\sqrt[6]{e}}{3e}\right) =$
33) $e^{-3\ln e + 3\ln 3} =$
34) $e^{\ln\frac{15}{e}} =$
35) $19\ln(e^e) =$

Exponential Equations and Logarithms

 Solve each equation for the unknown variable.

1) $3^{2n} = 27$
2) $5^r = 125$
3) $15^n = 85$
4) $8^{r+3} = 2$
5) $144^x = 12$
6) $7^{-3v-2} = 49$
7) $8^{2n} = 64$
8) $6^n = 1{,}296$
9) $\dfrac{15^{2a}}{3^{-a}} = 315$
10) $11 \times 11^{-v} = 1{,}331$
11) $3^{2n} = \dfrac{1}{81}$
12) $\left(\dfrac{1}{9}\right)^n = 81$
13) $256^{2x} = 4$
14) $9^{3-2x} = 9^{-x}$
15) $6^{-3x} = 6^{x-3}$
16) $2^{3n} = 32$
17) $12^{5x+3} = 12^{2x}$
18) $10^{2n} = 100$
19) $3^{-4k} = 243$
20) $3^r = 9^{-4r}$
21) $13^{x+3} = 13^{4x}$
22) $9^{3x} = 729$
23) $15 \times 15^{-v} = 225$
24) $\dfrac{81}{3^{-2m}} = 3^{-2m-1}$
25) $8^{-2n} \times 8^2 = 8^{-n}$
26) $\left(\dfrac{1}{9}\right)^{2n+1} \times \left(\dfrac{1}{9}\right)^{-n-10} = \left(\dfrac{1}{9}\right)^{-2n}$

 Solve each problem. (Round to the nearest whole number)

27) A substance decays 15% each day. After 11 days, there are 6 milligrams of the substance remaining. How many milligrams were there initially? _____

28) A culture of bacteria grows continuously. The culture doubles every 4 hours. If the initial number of bacteria is 13, how many bacteria will there be in 23 hours? _____

29) Bob plans to invest $12,000 at an annual rate of 6.5%. How much will Bob have in the account after six years if the balance is compounded quarterly? _____

30) Suppose you plan to invest $8,000 at an annual rate of 6%. How much will you have in the account after 4 years if the balance is compounded monthly? _____

Solving Logarithmic Equations

✎ **Find the value of the variables in each equation.**

1) $\log(x) + 8 = 4$

2) $-\log_3 4x = 5$

3) $\log(x) + 7 = 6$

4) $\log x - \log 7 = 4$

5) $\log x + \log 4 = 2$

6) $\log 4 + \log x = 3$

7) $\log x + \log 2 = \log 12$

8) $-3\log_3(x-2) = -15$

9) $\log 4x = \log(3x + 2)$

10) $\log(2k - 4) = \log(k - 5)$

11) $\log(5p - 2) = \log(-2p + 12)$

12) $-8 + \log_3(n + 3) = -8$

13) $\log_3(x + 5) = \log_3(x^2 + 8)$

14) $\log_9(v^2 + 24) = \log_9(-3v - 8)$

15) $\log(9 + 4b) = \log(7b^2 + 6b)$

16) $\log_9(x + 8) - \log_9 x = \log_9 7$

17) $\log_5 9 + \log_5 x^2 = \log_5 81$

18) $\log_6(x + 5) + \log_6 x = \log_6 24$

✎ **Find the value of x in each natural logarithm equation.**

19) $\ln 9 - \ln(3x + 9) = 3$

20) $\ln(x - 4) - \ln(x - 3) = \ln 4$

21) $\ln e^{27} - \ln(x + 3) = 3$

22) $\ln(2x - 6) - \ln(x - 12) = \ln 10$

23) $\ln 6x + \ln(x - 2) = \ln 3x$

24) $\ln(x - 3) - 2\ln(x - 3) = \ln 9$

25) $\ln(9x + 3) - \ln 5 = 6$

26) $\ln(x - 5) + \ln(x - 4) = \ln 2$

27) $\ln 8 + \ln(x + 4) = 10$

28)

29) $3\ln 3x - \ln(x + 9) = 3\ln 3x$

30) $\ln x^2 + \ln x^4 = \ln 1$

31) $\ln x^6 - \ln(x + 6) = 6\ln x$

32) $16\ln(x - 2) = 4\ln(x^2 - 4x + 4)$

33) $\ln(x^2 + 10) = \ln(3x + 8)$

34) $6\ln x - 6\ln(x + 3) = 12\ln(x^2)$

35) $\ln(2x - 3) - \ln(4x - 3) = \ln 4$

36) $\ln 3 + 9\ln(x + 2) = \ln 3$

37) $3\ln e^2 + \ln(3x - 2) = \ln 3 + 9$

Answers of Worksheets – Chapter 9

Rewriting Logarithms

1) $5^2 = 25$
2) $4^4 = 256$
3) $3^4 = 81$
4) $8^2 = 64$
5) $6^3 = 216$
6) $2^4 = 16$
7) $10^2 = 100$
8) $3^5 = 243$
9) $5^4 = 625$
10) $2^8 = 256$
11) $3^8 = 6,561$
12) $11^2 = 121$
13) $14^2 = 196$
14) $81^{\frac{1}{4}} = 3$
15) $27^{\frac{1}{3}} = 3$
16) $32^{\frac{1}{5}} = 2$
17) $512^{\frac{1}{3}} = 8$
18) $2^{-3} = \frac{1}{8}$
19) $2^{-4} = \frac{1}{16}$
20) $a^b = \frac{7}{3}$
21) $\log_{12} 144 = 2$
22) $\log_3 343 = 7$
23) $\log_4 1,024 = 5$
24) $\log_{15} 225 = 2$
25) $\log_5 625 = 4$
26) $\log_6 1,296 = 4$
27) $\log_2 512 = 9$
28) $\log_5 3,125 = 5$
29) $\log_4 \frac{1}{4,096} = -6$
30) $\log_3 \frac{1}{243} = -5$
31) $\log_{16} \frac{1}{256} = -2$
32) $\log_6 \frac{1}{216} = -3$
33) $\log_3 \frac{1}{19,683} = -9$
34) $\log_{21} \frac{1}{441} = -2$

Evaluating Logarithms

1) 7
2) 8
3) 3
4) 4
5) 5
6) 5
7) 2
8) −1
9) −2
10) −4
11) −3
12) −4
13) −2
14) $-\frac{1}{4}$
15) $\frac{1}{3}$
16) −6
17) −3
18) $-\frac{1}{3}$
19) $(4, 6)$
20) $(\frac{1}{4}, -6)$
21) $(7, 2)$
22) $(6, 5)$
23) $(4, 2)$
24) $(7, -4)$
25) $(3, 3)$

Properties of Logarithms

1) $\log 9 + \log 4$
2) $\log 6 + \log 3$
3) $\log 2 + \log 8$
4) $\log 8 - \log 7$
5) $\log 9 - \log 5$
6) $3 \log 4 - 3 \log 11$
7) $\log 9 + 3 \log 4$
8) $2\log 7 - 2 \log 3$
9) $4 \log 5 - \log 9$
10) $7 \log x + 7 \log y$

11) $2\log x + \log y + 5\log z$
12) $8\log u - \log v$
13) $\log x - 4\log y$
14) $\log \frac{7}{12}$
15) $\log(8 \times 3)$
16) $\log \frac{2^4}{5^7}$
17) $\log \frac{4^6}{9^5}$
18) $\log \frac{8^3}{17}$
19) $\log \frac{3^8}{2^6}$

20) $\log \frac{11}{5^2}$
21) $\log(6^4 \times 9^3)$
22) $\log(5^4 \times 13^5)$
23) $\log_5 (a^7 b^{16})$
24) $\log_6 \frac{x^5}{y^7}$
25) $\log_5 \frac{u}{v^9}$
26) $\log_3(u^8 \times v^{21})$
27) $\log_7 \frac{u^{26}}{v^{15}}$

Natural Logarithms

1) $x = \ln 9$
2) $x = \ln 36, x = 2\ln(6)$
3) $x = \ln 49, x = 2\ln(7)$
4) $x = e^3$
5) $x = e^{e^3}$
6) $x = \ln 11$
7) $x = \frac{e-9}{5}$
8) $x = \frac{e^3-3}{7}$
9) $x = \frac{e^4-5}{8}$
10) $x = \sqrt[3]{e}$
11) $x = \frac{ee^4}{6}$
12) $x = 28$

13) $x = 216$
14) 3
15) 10
16) 4
17) 15
18) 13
19) 12
20) 19
21) 25
22) 81
23) $\frac{1}{6}$
24) $\frac{4}{9}$
25) $\frac{25}{16e^2}$

26) 12
27) -9
28) 750
29) $\frac{13}{e}$
30) 0
31) 12
32) -5.8
33) $27e^{-3} = \frac{27}{e^3}$
34) $\frac{15}{e}$
35) $19e$

Exponential Equations and Logarithms

1) $\frac{3}{2}$
2) 3
3) 1.64
4) $\frac{-8}{3}$
5) $\frac{1}{2}$
6) $-\frac{4}{3}$
7) 1
8) 0.883
9) -2
10) -2

11) -2
12) $\frac{1}{8}$
13) 3
14) $\frac{3}{4}$
15) $\frac{5}{3}$
16) -1

17) 1
18) $-\frac{5}{4}$
19) 0
20) 1
21) 1
22) -1

23) -1.25
24) 2
25) 3
26) 35.9
27) 699.6
28) $\$17,668.3$
29) $\$10,163.9$

Solving Logarithmic Equations

1) $\{\frac{1}{10,000}\}$
2) $\{\frac{1}{972}\}$
3) $\{\frac{1}{10}\}$
4) $\{70,000\}$
5) $\{25\}$
6) $\{250\}$
7) $\{6\}$
8) $\{245\}$
9) $\{2\}$
10) No Solution
11) $\{2\}$
12) $\{-2\}$
13) No Solution

14) No Solution
15) $\{1, -\frac{9}{7}\}$
16) $\{\frac{4}{3}\}$
17) $\{3, -3\}$
18) $\{3\}$
19) $x = \frac{3-3e^3}{e^3}$
20) No Solution
21) $e^{24} - 3$
22) $\{\frac{57}{4}\}$
23) $\{\frac{5}{2}\}$
24) $\{\frac{28}{9}\}$

25) $x = \frac{5e^6 - 3}{9}$
26) $x = 6$
27) $x = \frac{e^{10} - 32}{8}$
28) No Solution
29) $\{1, -1\}$
30) No Solution
31) $x = 3$
32) $\{1, 2\}$
33) $\{0.64951 \dots\}$
34) No Solution
35) $\{-1\}$
36) $x = \frac{3e^3 + 2}{3}$

Chapter 10:
Conic Sections

Topics that you'll learn in this chapter:

- ✓ Equation of a Parabola
- ✓ Focus, Vertex, and Directrix of a Parabola
- ✓ Standard Form of a Circle
- ✓ Standard Equation of an Ellipse
- ✓ Hyperbola in Standard Form
- ✓ Conic Sections in Standard Form

He who takes nature for his guide, is not easily beaten out of his argument.
— Thomas Paine

Equation of a Parabola

✎ **Write the equation of the following parabolas.**

1) Vertex (0, 0) and Focus (0, 2)

2) Vertex (3, 2) and Focus (3, 4)

3) Vertex (1, 1) and Focus (1, 6)

4) Vertex (− 1, 2) and Focus (− 1, 5)

5) Vertex (2, 2) and Focus (2, 6)

6) Vertex (0, 1) and Focus (0, 2)

7) Vertex (2, 1) and Focus (4, 1)

8) Vertex (5, 0) and Focus (9, 0)

9) Vertex (− 2, 4) and Focus (2, 4)

10) Vertex (− 4, 2) and Focus (0, 2)

Focus, Vertex, and Directrix of a Parabola

✏ **Use the information provided to write the vertex form equation of each parabola.**

1) $y = x^2 + 8x$

2) $y = x^2 - 6x + 5$

3) $y + 6 = (x + 3)^2$

4) $y = x^2 + 10x + 33$

5) $y = (x + 5)(x + 4)$

6) $\frac{1}{2}(y + 4) = (x - 7)^2$

7) $162 + 731 = -y - 9x^2$

8) $y = x^2 + 16x + 71$

9) Focus: $(-\frac{63}{8}, -7)$, Directrix: $x = -\frac{65}{8}$

10) Focus: $(\frac{107}{12}, -7)$, Directrix: $x = \frac{109}{12}$

11) Opens down or up, and passes through $(-6, -7), (-11, -2)$, and $(-8, 1)$

12) Opens down or up, and passes through $(11, 15), (7, 7)$, and $(4, 22)$

Algebra 2 Workbook

Standard Form of a Circle

✎ **Write the standard form equation of each circle.**

1) $x^2 + y^2 - 8x - 6y + 21 = 0$

2) $y^2 + 2x + x^2 = 24y - 120$

3) $x^2 + y^2 - 2y - 15 = 0$

4) $8x + x^2 - 2y = 64 - y^2$

5) Center: (–5, –6), Radius: 9

6) Center: (–9, –12), Radius: 4

7) Center: (–12, –5), Area: 4π

8) Center: (–11, –14), Area: 16π

9) Center: (–3, 2), Circumference: 2π

10) Center: (15, 14), Circumference: $2\pi\sqrt{15}$

✎ **Identify the center and radius of each. Then sketch the graph.**

11) $(x - 2)^2 + (y + 5)^2 = 10$

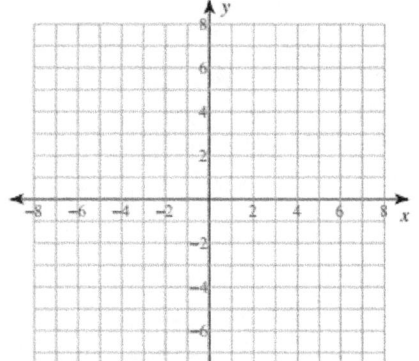

12) $x^2 + (y - 1)^2 = 4$

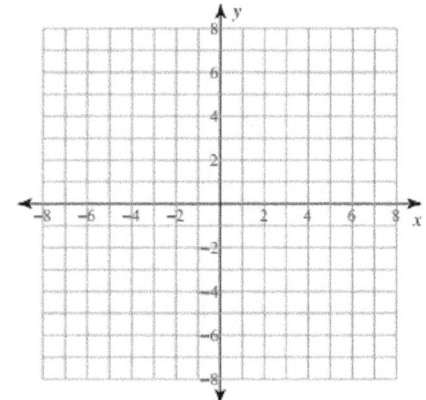

13) $(x - 2)^2 + (y + 6)^2 = 9$

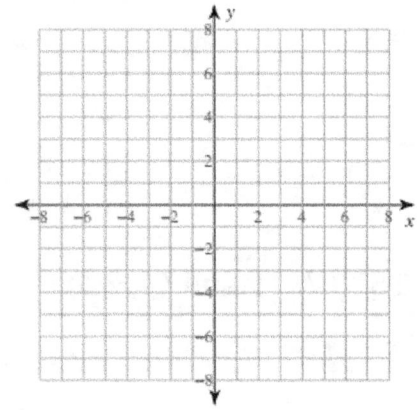

14) $(x + 14)^2 + (y - 5)^2 = 16$

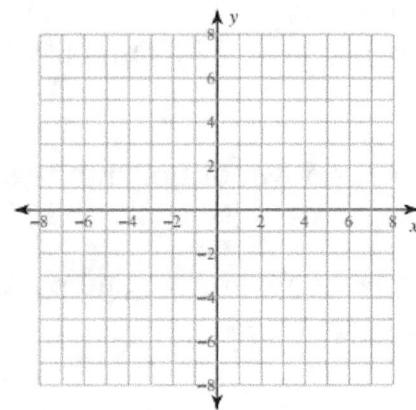

WWW.MathNotion.Com

Equation of Each Ellipse

✍ **Use the information provided to write the standard form equation of each ellipse.**

1) Foci: $(2\sqrt{3}, 0)$, $(-2\sqrt{3}, 0)$; Co–vertices: $(0, 2)$. $(0, -2)$

2) Vertices: $(0, 6)$, $(0, -6)$; Co–vertices: $(3, 0)$. $(-3, 0)$

3) Vertices: $(4, 3)$, $(4, -7)$; Co–vertices: $(1, -2)$. $(7, -2)$

4) Foci: $(\sqrt{17}, 0)$, $(-\sqrt{17}, 0)$; Co–vertices: $(9, 0)$. $(-9, 0)$

5) Foci: $(-7, 5 + \sqrt{13})$, $(-7, 5 - \sqrt{13})$; Co–vertices: $(-1, 5)$. $(-13, 5)$

6) Vertices: $(5, 1)$, $(-1, 1)$; Co–vertices: $(2, 3)$. $(2, -1)$

7) Vertices: $(12, 0)$, $(-12, 0)$; Co–vertices: $(2\sqrt{11}, 0)$. $(-2\sqrt{11}, 0)$

8) Vertices: $(7 + 2\sqrt{35}, -4)$, $(7 - 2\sqrt{35}, -4)$; Co–vertices: $(7, -2)$. $(7, -6)$

9) Center: $(4, 8)$; Vertex: $(4, 8 -\sqrt{170})$; Co–vertex: $(4 - \sqrt{15}, 8)$

10) Center: $(7, -10)$; Vertex: $(-6, -10)$; Co–vertex: $(7, -17)$

✍ **Identify the vertices, co–vertices, foci**

11) $\dfrac{x^2}{169} + \dfrac{y^2}{64} = 1$

12) $\dfrac{x^2}{95} + \dfrac{y^2}{30} = 1$

13) $\dfrac{x^2}{36} + \dfrac{y^2}{16} = 1$

14) $\dfrac{x^2}{49} + \dfrac{y^2}{169} = 1$

15) $\dfrac{(x+5)^2}{81} + \dfrac{(y-1)^2}{144} = 1$

16) $\dfrac{(x-3)^2}{49} + \dfrac{(y-9)^2}{4} = 1$

17) $\dfrac{x^2}{64} + \dfrac{(y-8)^2}{9} = 1$

18) $\dfrac{x^2}{64} + \dfrac{(y-6)^2}{121} = 1$

Hyperbola in Standard Form

✏ **Use the information provided to write the standard form equation of each hyperbola.**

1) $-2x^2 + 3y^2 + 4x - 60y + 268 = 0$

2) $-x^2 + y^2 - 18x - 14y - 132 = 0$

3) $-16x^2 + 9y^2 + 32x + 144y - 16 = 0$

4) $9x^2 - 4y^2 - 90x + 32y - 163 = 0$

5) Vertices: (8, 14), (8, −10), Conjugate Axis is 6 units long

6) Vertices: (7, 4), (7, −24), Distance from Center to Focus = $7\sqrt{5}$

7) Vertices: (−5, 22), (−5, −4), Distance from Center to Focus = $\sqrt{218}$

8) Vertices: (0, −1), (−20, −1), Asymptotes: $y = x + 9$, $y = -x - 11$

9) Foci: $(-9, -5 + 9\sqrt{2})$, $(-9, -5 - 9\sqrt{2})$; Conjugate Axis is 18 units long

10) Foci: $(8, -5 + \sqrt{53})$, $(8, -5 - \sqrt{53})$,

    Endpoints of Conjugate Axis: (15, −5), (1, −5)

✏ **Identify the vertices, foci, and direction of opening of each.**

11) $\dfrac{y^2}{25} - \dfrac{x^2}{16} = 1$

12) $\dfrac{x^2}{121} - \dfrac{y^2}{36} = 1$

13) $\dfrac{x^2}{121} - \dfrac{y^2}{81} = 1$

14) $\dfrac{x^2}{81} - \dfrac{y^2}{4} = 1$

15) $\dfrac{(x+2)^2}{169} - \dfrac{(y+8)^2}{4} = 1$

16) $\dfrac{(y+8)^2}{36} - \dfrac{(y+2)^2}{25} =$

Conic Sections in Standard Form

✎ **Classify each conic section and write its equation in standard form.**

1) $x^2 - 4y^2 + 6x - 8y + 1 = 0$

2) $3x^2 + 3x + y + 79 = 0$

3) $x^2 + y^2 + 4x - 2y - 18 = 0$

4) $-y^2 + x + 8y - 17 = 0$

5) $49x^2 + 9y^2 + 392x + 343 = 0$

6) $-9x^2 + y^2 - 72x - 153 = 0$

7) $-2y^2 + x - 20y - 49 = 0$

8) $-x^2 + 10x + y - 21 = 0$

✎ **Classify each conic section. (Not in Standard Form)**

9) $x^2 + y^2 - 8x + 8y - 4 = 0$

10) $y = 6x^2 - 60x + 149$

11) $x^2 - 4x + 4y^2 - 32y + 32 = 0$

12) $x^2 - 2x - 36y^2 - 360y - 935 = 0$

13) $y = 6x^2 - 60x + 149$

14) $x^2 + y^2 - 8x + 8y - 4 = 0$

15) $x^2 + y^2 + 6x + 10y + 33 = 0$

16) $x^2 - 4x - 36y^2 + 288y - 608 = 0$

17) $9x^2 + 4y^2 + 16y - 128 = 0$

18) $x^2 + 8x - 25y^2 + 50y - 34 = 0$

19) $y = 6x^2 + 60x + 155$

20) $4x^2 + 9y^2 - 54y + 45 = 0$

21) $-9x^2 - 54x + 4y^2 - 40y - 125 = 0$

22) $x^2 - 4x + 4y^2 - 32y + 32 = 0$

Answers of Worksheets – Chapter 10

Equation of a Parabola

1) $x^2 = 8y$
2) $(x - 3)^2 = 8(y - 2)$
3) $(x - 1)^2 = 20(y - 1)$
4) $(x + 1)^2 = 12(y - 2)$
5) $(x - 2)^2 = 8(y - 2)$
6) $x^2 = 8(y - 1)$
7) $(y - 1)^2 = 8(x - 2)$
8) $(y - 1)^2 = 8(x - 2)$
9) $(y - 4)^2 = 16(x + 2)$
10) $(y + 4)^2 = 16x$

Focus, Vertex, and the Directrix of a Parabola

1) $y = (x + 4)^2 - 16$
2) $y = (x - 3)^2 - 4$
3) $y = (x + 3)^2 - 6$
4) $y = (x + 5)^2 + 8$
5) $y = (x + \frac{9}{2})^2 - \frac{1}{4}$
6) $y = 2(x - 7)^2 - 4$
7) $y = -9(x + 9)^2 - 2$
8) $y = (x + 8)^2 + 7$
9) $x = 2(y + 7)^2 - 8$
10) $x = -3(y + 7)^2 + 9$
11) $y = -(x + 9)^2 + 2$
12) $y = (x - 8)^2 + 6$

Standard Form of a Circle

1) $(x - 4)^2 + (y - 3)^2 = 4$
2) $(x + 1)^2 + (y - 12)^2 = 25$
3) $x^2 + (y - 1)^2 = 16$
4) $(x + 4)^2 + (y - 1)^2 = 81$
5) $(x + 5)^2 + (y + 6)^2 = 81$
6) $(x + 9)^2 + (y + 12)^2 = 16$
7) $(x + 12)^2 + (y + 5)^2 = 4$
8) $(x + 11)^2 + (y + 14)^2 = 16$
9) $(x + 3)^2 + (y - 2)^2 = 1$
10) $(x - 15)^2 + (y - 14)^2 = 15$

11) Center: $(2, -5)$, Radius: $\sqrt{10}$

12) Center: $(0, 1)$, Radius: $2\sqrt{26}$

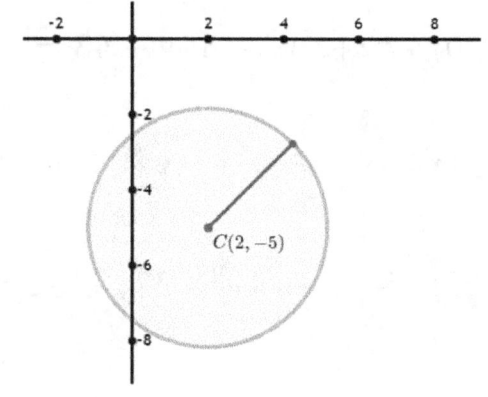
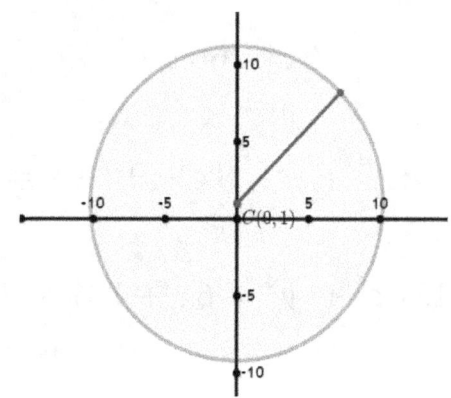

13) Center: (2, – 6), Radius: 3

14) Center: (–14, –5), Radius: 4

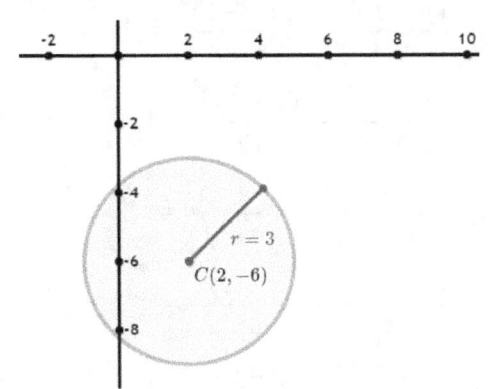
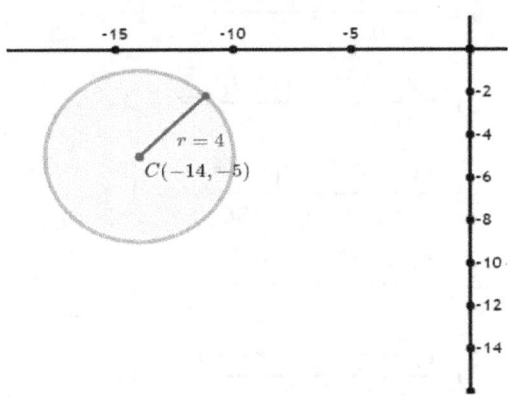

Equation of Each Ellipse

1) $\dfrac{x^2}{16} + \dfrac{y^2}{4} = 1$

2) $\dfrac{x^2}{9} + \dfrac{y^2}{36} = 1$

3) $\dfrac{(x-4)^2}{9} + \dfrac{(y+2)^2}{25} = 1$

4) $\dfrac{x^2}{81} + \dfrac{y^2}{64} = 1$

5) $\dfrac{(x+7)^2}{36} + \dfrac{(y-5)^2}{49} = 1$

6) $\dfrac{(x-2)^2}{9} + \dfrac{(y-1)^2}{4} = 1$

7) $\dfrac{x^2}{144} + \dfrac{y^2}{100} = 1$

8) $\dfrac{(x-7)^2}{144} + \dfrac{(y+5)^2}{4} = 1$

9) $\dfrac{(x-4)^2}{15} + \dfrac{(y-8)^2}{170} = 1$

10) $\dfrac{(x-7)^2}{169} + \dfrac{(y+10)^2}{49} = 1$

11) Vertices: (13, 0), (–13, 0); Co–vertices: (0, 8), (0, –8); Foci: ($\sqrt{105}$, 0), (–$\sqrt{105}$, 0)

12) Vertices: ($\sqrt{95}$, 0), (–$\sqrt{95}$, 0); Co–vertices: (0, $\sqrt{30}$), (0, –$\sqrt{30}$); Foci: ($\sqrt{65}$, 0), (–$\sqrt{65}$, 0)

13) Vertices: (6, 0), (–6, 0); Co–vertices: (0, 4), (0, –4); Foci: ($2\sqrt{5}$, 0), (–$2\sqrt{5}$, 0)

14) Vertices: (0, 13), (0, –13); Co–vertices: (7, 0), (–7, 0); Foci: (0, $2\sqrt{30}$), (0, –$2\sqrt{30}$)

15) Vertices: (–5, 13), (–5, –11); Co–vertices: (4, 1), (–14, 1);

    Foci: (–5, $1 + 3\sqrt{7}$), (–5, $1 - 3\sqrt{7}$)

16) Vertices: (10, 9), (–4, 9); Co–vertices: (3, 11), (3, 7);

    Foci: ($3 + 3\sqrt{5}$, 9), ($3 - 3\sqrt{5}$, 9)

17) Vertices: (8, 8), (–8, 8); Co–vertices: (0, 11), (0, 5);

    Foci: ($\sqrt{55}$, 8), (–$\sqrt{55}$, 8)

18) Vertices: (0, 17), (0, –5); Co–vertices: (8, 6), (–8, 6);

    Foci: (0, $6 + \sqrt{57}$), (0, $6 - \sqrt{57}$)

Algebra 2 Workbook

Hyperbola in Standard Form

1) $\dfrac{(y-10)^2}{10} - \dfrac{(x-1)^2}{15} = 1$

2) $\dfrac{(y-7)^2}{100} - \dfrac{(x+9)^2}{100} = 1$

3) $\dfrac{(y+8)^2}{64} - \dfrac{(x-1)^2}{36} = 1$

4) $\dfrac{(x-5)^2}{36} - \dfrac{(y-4)^2}{81} = 1$

5) $\dfrac{(y-2)^2}{144} - \dfrac{(x-8)^2}{9} = 1$

6) $\dfrac{(y+10)^2}{196} - \dfrac{(x-7)^2}{49} = 1$

7) $\dfrac{(y-9)^2}{196} - \dfrac{(x+5)^2}{49} = 1$

8) $\dfrac{(x+10)^2}{100} - \dfrac{(y+1)^2}{100} = 1$

9) $\dfrac{(y+5)^2}{81} - \dfrac{(x+9)^2}{81} = 1$

10) $\dfrac{(y+5)^2}{4} - \dfrac{(x-8)^2}{49} = 1$

11) Vertices: (0, 5), (0, –5); Foci: (0, $\sqrt{41}$), (0, –$\sqrt{41}$); Opens up/down

12) Vertices: (11, 0), (–11, 0); Foci: ($\sqrt{157}$, 0), (–$\sqrt{157}$, 0); Opens left/right

13) Vertices: (11, 0), (–11, 0); Foci: ($\sqrt{202}$, 0), (–$\sqrt{202}$, 0); Opens left/right

14) Vertices: (9, 0), (–9, 0); Foci: ($\sqrt{85}$, 0), (–$\sqrt{85}$, 0); Opens left/right

15) Vertices: (11, –8), (–15, –8); Foci: (–2 + $\sqrt{173}$, –8), (–2 – $\sqrt{173}$, –8); Opens left/right

16) Vertices: (–2, –2), (–2, –14); Foci: (–2, –8 + $\sqrt{61}$), (–2, –8 – $\sqrt{61}$); Opens up/down

Conic Sections in Standard Form

1) Hyperbola, $\dfrac{(x+3)^2}{4} - (y+1)^2 = 1$

2) Parabola, $y = -3(x+5)^2 - 4$

3) Circle, $(x+2)^2 + (y-1)^2 = 23$

4) Parabola, $x = (y-4)^2 + 1$

5) Ellipse, $\dfrac{(x+4)^2}{9} + \dfrac{y^2}{49} = 1$

6) Hyperbola, $\dfrac{y^2}{9} - (x+4)^2 = 1$

7) Parabola, $x = 2(y+5)^2 - 1$

8) Parabola, $y = (x-5)^2 - 4$

9) Circle

10) Parabola

11) Ellipse

12) Hyperbola

13) Parabola

14) Circle

15) Circle

16) Hyperbola

17) Ellipse

18) Hyperbola

19) Parabola

20) Ellipse

21) Hyperbola

22) Ellipse

Chapter 11:
Trigonometric Functions

Topics that you will practice in this chapter:

- ✓ Trig ratios of General Angles
- ✓ Sketch Each Angle in Standard Position
- ✓ Finding Co-Terminal Angles and Reference Angles
- ✓ Angles in Radians
- ✓ Angles in Degrees
- ✓ Evaluating Each Trigonometric Expression
- ✓ Missing Sides and Angles of a Right Triangle
- ✓ Arc Length and Sector Area

Mathematics is like checkers in being suitable for the young, not too difficult, amusing, and without peril to the state. — Plato

Trig ratios of General Angles

✎ **Evaluate.**

1) $\sin -135° = $ _____

2) $\sin 300° = $ _____

3) $\cos -390° = $ _____

4) $\cos 240° = $ _____

5) $\sin 390° = $ _____

6) $\sin -330° = $ _____

7) $\tan 120° = $ _____

8) $\cot 150° = $ _____

9) $\tan 210° = $ _____

10) $\cot 225° = $ _____

11) $\sec 330° = $ _____

12) $\csc 450° = $ _____

13) $\cot -135° = $ _____

14) $\sec 360° = $ _____

15) $\cos -450° = $ _____

16) $\sec 150° = $ _____

17) $\csc 360° = $ _____

18) $\cot -120° = $ _____

✎ **Find the exact value of each trigonometric function. Some may be undefined.**

19) $\sec 5\pi = $ _____

20) $\tan -\frac{5\pi}{2} = $ _____

21) $\cos \frac{4\pi}{2} = $ _____

22) $\cot \frac{20\pi}{6} = $ _____

23) $\sec -\frac{21\pi}{6} = $ _____

24) $\sec \frac{4\pi}{3} = $ _____

25) $\csc \frac{17\pi}{3} = $ _____

26) $\cot \frac{3\pi}{4} = $ _____

27) $\csc -\frac{7\pi}{6} = $ _____

28) $\cot \frac{2\pi}{3} = $ _____

Algebra 2 Workbook

Sketch Each Angle in Standard Position

✍ **Draw each angle with the given measure in standard position.**

1) 390°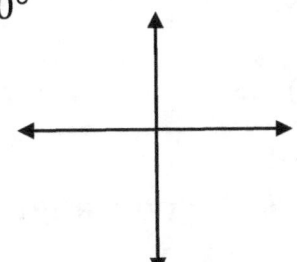

2) 750°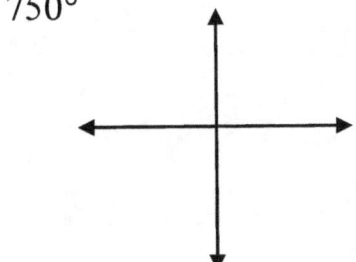

3) −300°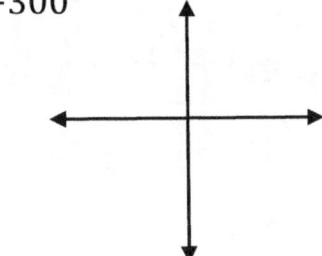

4) 1,100°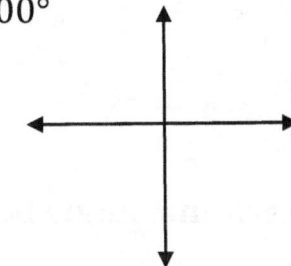

5) $-\frac{11\pi}{6}$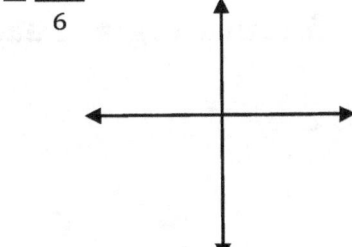

6) $\frac{23\pi}{6}$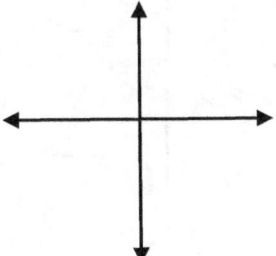

Finding Co-terminal Angles and Reference Angles

✎ **Find a conterminal angle between 0° and 360° for each angle provided.**

1) $-150° =$

2) $-270° =$

3) $-315° =$

4) $-810° =$

✎ **Find a conterminal angle between 0 and 2π for each given angle.**

5) $\dfrac{12\pi}{5} =$

6) $-\dfrac{19\pi}{8} =$

7) $-\dfrac{25\pi}{11} =$

8) $\dfrac{21\pi}{4} =$

✎ **Find the reference angle of each angle.**

9)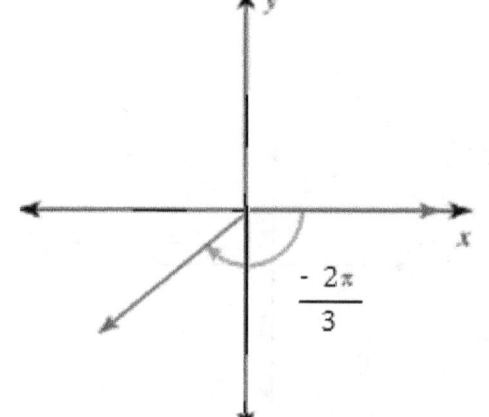
$-\dfrac{2\pi}{3}$

10)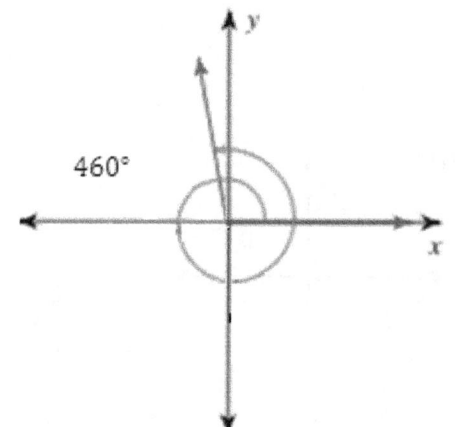
460°

Angles and Angle Measure

✎ **Convert each degree measure into radians.**

1) $225° =$ ___

2) $432° =$ ___

3) $840° =$ ___

4) $576° =$ ___

5) $420° =$ ___

6) $810° =$ ___

7) $-270° =$ ___

8) $855° =$ ___

9) $330° =$ ___

10) $288° =$ ___

11) $100° =$ ___

12) $420° =$ ___

13) $-150° =$ ___

14) $-280° =$ ___

15) $-260° =$ ___

16) $630° =$ ___

17) $-960° =$ ___

18) $864° =$ ___

19) $-144° =$ ___

20) $1,170° =$ ___

21) $810° =$ ___

✎ **Convert each radian measure into degrees.**

22) $\frac{\pi}{10} =$

23) $\frac{2\pi}{18} =$

24) $\frac{8\pi}{9} =$

25) $\frac{6\pi}{15} =$

26) $-\frac{17\pi}{10} =$

27) $\frac{5\pi}{18} =$

28) $-\frac{8\pi}{15} =$

29) $\frac{15\pi}{18} =$

30) $\frac{7\pi}{60} =$

31) $\frac{18\pi}{20} =$

32) $-\frac{3\pi}{18} =$

33) $\frac{13\pi}{90} =$

34) $-\frac{\pi}{9} =$

35) $\frac{12\pi}{30} =$

36) $-\frac{8\pi}{72} =$

37) $\frac{7\pi}{6} =$

38) $-\frac{8\pi}{40} =$

39) $-\frac{11\pi}{90} =$

Algebra 2 Workbook

Evaluating Trigonometric Functions

✍ **Find the exact value of each trigonometric function.**

1) $\cos 510° = $ _____

2) $\tan \dfrac{5\pi}{3} = $ _____

3) $\tan -\dfrac{11\pi}{2} = $ _____

4) $\cot -\dfrac{7\pi}{3} = $ _____

5) $\cos -\dfrac{19\pi}{4} = $ _____

6) $\cos -315° = $ _____

7) $\sin 405° = $ _____

8) $\tan 450° = $ _____

9) $\cot -480° = $ _____

10) $\tan 495° = $ _____

11) $\cot 810° = $ _____

12) $\sin -540° = $ _____

13) $\cot -585° = $ _____

✍ **Use the given point on the terminal side of angle θ to find the value of the trigonometric function indicated.**

14) $\sin\theta;\ (-6, 8)$

15) $\cos\theta;\ (-6, 8)$

16) $\cot\theta;\ (-2, -6)$

17) $\cos\theta;\ (10, 24)$

18) $\sin\theta;\ (9, -9)$

19) $\tan\theta;\ (-6, -\sqrt{12})$

Missing Sides and Angles of a Right Triangle

✎ Find the value of each trigonometric ratio as fractions in their simplest form.

1) $\tan x$

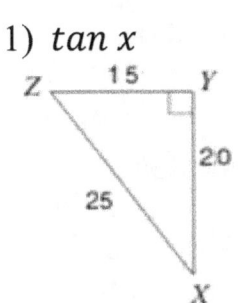

2) $\sin A$

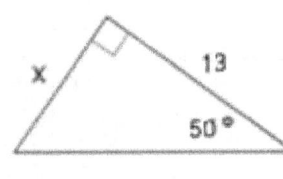

✎ Find the missing sides. Round answers to the nearest tenth.

3)

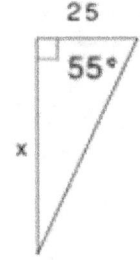

4)

5)

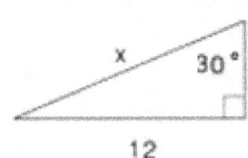

6)

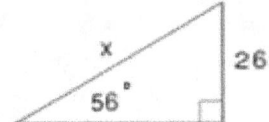

Arc Length and Sector Area

📝 **Find the length of each arc. Round your answers to the nearest tenth.**

($\pi = 3.14$)

1) $r = 40$ cm, $\theta = 90°$

2) $r = 16$ ft, $\theta = 55°$

3) $r = 24$ ft, $\theta = 100°$

4) $r = 18$ m, $\theta = 85°$

📝 **Find area of each sector. Do *not* round. Round your answers to the nearest tenth.** ($\pi = 3.14$)

5)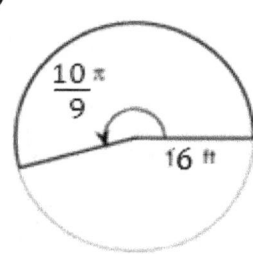

7)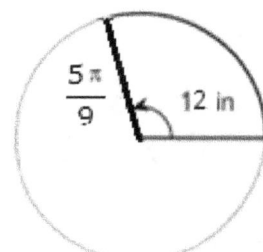

6)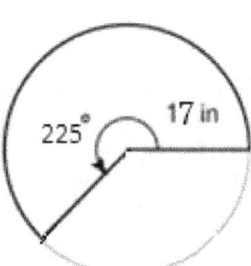

8)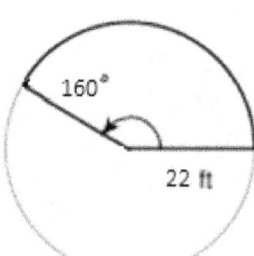

Answers of Worksheets – Chapter 11

Trig Ratios of General Angles

1) $-\frac{\sqrt{2}}{2}$

2) $-\frac{\sqrt{3}}{2}$

3) $\frac{\sqrt{3}}{2}$

4) $-\frac{1}{2}$

5) $\frac{1}{2}$

6) $\frac{1}{2}$

7) $-\sqrt{3}$

8) $-\sqrt{3}$

9) $\frac{\sqrt{3}}{3}$

10) 1

11) $\frac{2\sqrt{3}}{3}$

12) 1

13) 1

14) 1

15) 0

16) $-\frac{2\sqrt{3}}{3}$

17) Undefined

18) $\frac{\sqrt{3}}{3}$

19) -1

20) Undefined

21) 1

22) $\frac{\sqrt{3}}{3}$

23) Undefined

24) -2

25) $-\frac{2\sqrt{3}}{3}$

26) -1

27) 2

28) $-\frac{\sqrt{3}}{3}$

Sketch Each Angle in Standard Position

1) $390°$

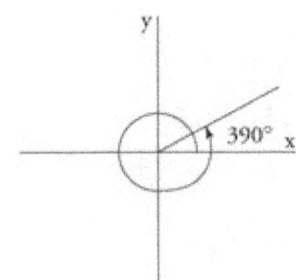

2) $750°$

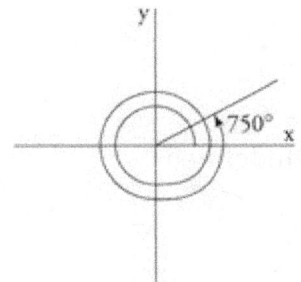

3) $-300°$

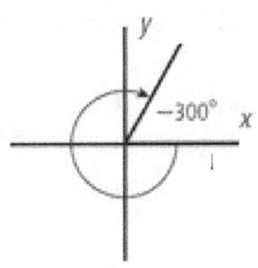

4) $1,110°$

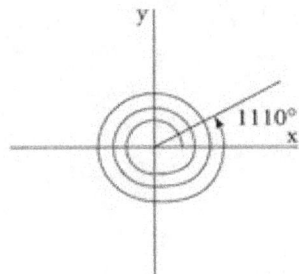

5) $-\frac{11\pi}{6} = -330°$

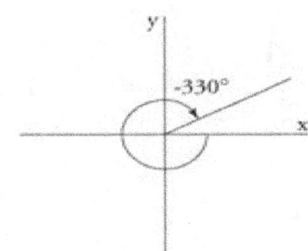

6) $\frac{23\pi}{6} = -690°$

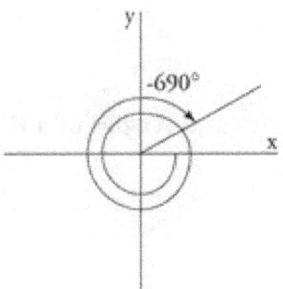

Finding Co–Terminal Angles and Reference Angles

1) $210°$

2) $90°$

3) $45°$

4) $270°$

5) $\frac{2\pi}{5}$

6) $\frac{13\pi}{8}$

7) $\frac{19\pi}{11}$

8) $\frac{5\pi}{4}$

9) $\frac{4\pi}{3}$

10) $100°$

Angles and Angle Measure

1) $\frac{5\pi}{4}$
2) $\frac{12\pi}{5}$
3) $\frac{14\pi}{3}$
4) $\frac{16\pi}{5}$
5) $\frac{7\pi}{3}$
6) $\frac{9\pi}{2}$
7) $-\frac{3\pi}{2}$
8) $\frac{19\pi}{4}$
9) $\frac{11\pi}{6}$
10) $\frac{8\pi}{5}$
11) $\frac{5\pi}{9}$
12) $\frac{21\pi}{9}$
13) $-\frac{5}{6}\pi$
14) $-\frac{14\pi}{9}$
15) $-\frac{13\pi}{9}$
16) $\frac{7\pi}{2}$
17) $-\frac{16\pi}{3}$
18) $\frac{24\pi}{5}$
19) $-\frac{4\pi}{5}$
20) $\frac{13\pi}{2}$
21) $\frac{27\pi}{6}$
22) 18°
23) 20°
24) 160°
25) 72°
26) −306°
27) 50°
28) −96°
29) 150°
30) 21°
31) 162°
32) −30°
33) 26°
34) −20°
35) 72°
36) −20°
37) 210°
38) −36°
39) −22°

Evaluating Each Trigonometric Functions

1) $-\frac{\sqrt{3}}{2}$
2) $-\sqrt{3}$
3) Undefined
4) $-\frac{\sqrt{3}}{3}$
5) $-\frac{\sqrt{2}}{2}$
6) $\frac{\sqrt{2}}{2}$
7) $\frac{\sqrt{2}}{2}$
8) Undefined
9) $\frac{\sqrt{3}}{3}$
10) −1
11) 0
12) 0
13) −1
14) 0.6
15) −0.8
16) 3
17) $\frac{10}{26}$
18) $-\frac{\sqrt{2}}{2}$
19) $\sqrt{3}$

Missing Sides and Angles of a Right Triangle

1) $\frac{3}{4}$
2) $\frac{12}{13}$
3) 35.7
4) 10
5) 24
6) 31.4

Arc Length and Sector Area

1) 62.8 cm
2) 15.4 ft
3) 41.9 ft
4) 26.7 m
5) 446.6 ft^2
6) 567.2 in^2
7) 125.6 in^2
8) 675.4 ft^2

Chapter 12:
Statistics and Probability

Topics that you will practice in this chapter:

- ✓ Mean and Median
- ✓ Mode and Range
- ✓ Probability Problems
- ✓ Factorials
- ✓ Combinations and Permutation

Mathematics is no more computation than typing is literature.

− John Allen Paulos

Mean and Median

✎ Find Mean and Median of the Given Data.

1) 8, 9, 19, 3, 4

2) 11, 7, 35, 10, 17, 32, 24

3) 38, 9, 15, 17, 13

4) 50, 19, 2, 18, 6, 7

5) 25, 27, 13, 16, 6, 13, 54

6) 24, 364, 42, 57, 6, 68

7) 89, 98, 65, 45, 3, 4, 30, 42

8) 34, 15, 15, 17, 22, 29, 15

9) 2, 5, 10, 45, 8, 13, 35, 6

10) 20, 22, 18, 7, 2, 17, 44, 53

11) 33, 52, 81, 9, 45, 31

12) 19, 74, 51, 8, 12, 15, 9, 14

✎ Calculate.

13) In a javelin throw competition, five athletics score 45, 33, 53, 46 and 19 meters. What are their Mean and Median? _____

14) Eva went to shop and bought 5 apples, 9 peaches, 4 bananas, 7 pineapples and 8 melons. What are the Mean and Median of her purchase? _____

15) Bob has 19 black pen, 15 red pen, 27 green pens, 21 blue pens and one boxes of yellow pens. If the Mean and Median are 19 respectively, what is the number of yellow pens in box? _____

Mode and Range

✍ **Find Mode and Rage of the Given Data.**

1) 7, 4, 18, 9, 9, 3
   Mode: _____   Range: _____

2) 8, 8, 15, 14, 8, 5, 6, 18
   Mode: _____   Range: _____

3) 4, 4, 4, 15, 19, 24, 31, 5, 4
   Mode: _____   Range: _____

4) 10, 10, 9, 17, 14, 8, 20, 4
   Mode: _____   Range: _____

5) 5, 11, 3, 4, 3, 3
   Mode: _____   Range: _____

6) 13, 7, 7, 7, 7, 4, 12, 25, 8, 3
   Mode: _____   Range: _____

7) 1, 7, 9, 9, 24, 24, 24, 20, 34, 35
   Mode: _____   Range: _____

8) 9, 4, 7, 13, 13, 13, 9, 8, 15
   Mode: _____   Range: _____

9) 8, 8, 8, 5, 8, 7, 17, 16, 3, 9
   Mode: _____   Range: _____

10) 34, 34, 32, 14, 6, 14, 9, 14
    Mode: _____   Range: _____

11) 8, 8, 6, 8, 18, 10, 16, 15
    Mode: _____   Range: _____

12) 12, 12, 7, 11, 14, 12, 33, 5
    Mode: _____   Range: _____

✍ **Calculate.**

13) A stationery sold 21 pencils, 42 red pens, 25 blue pens, 26 notebooks, 21 erasers, 28 rulers and 27 color pencils. What are the Mode and Range for the stationery sells?

   Mode: _____   Range: _____

14) In an English test, eight students score 19, 10, 10, 17, 35, 35, 14 and 10. What are their Mode and Range? _____

15) What is the range of the first 6 odd numbers greater than 8?
   _____

Probability Problems

✎ **Calculate.**

1) A number is chosen at random from 1 to 20. Find the probability of selecting number 8 or smaller numbers. _____

2) Bag A contains 16 red marbles and 6 green marbles. Bag B contains 12 black marbles and 18 orange marbles. What is the probability of selecting a green marble at random from bag A? What is the probability of selecting a black marble at random from Bag B? _____

3) A number is chosen at random from 1 to 25. What is the probability of selecting multiples of 5? _____

4) A card is chosen from a well-shuffled deck of 52 cards. What is the probability that the card will be a queen? _____

5) A number is chosen at random from 1 to 15. What is the probability of selecting a multiple of 4? _____

A spinner, numbered 1–8, is spun once. What is the probability of spinning …?

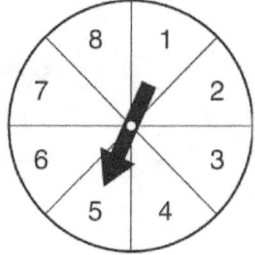

6) an Odd number? _____   7) a multiple of 2? ____

8) a multiple of 5? ____   9) number 10? _____

Factorials

✎ **Determine the value for each expression.**

1) $6! + 1! =$

2) $5! + 2! =$

3) $(4!)^2 =$

4) $6! - 3! =$

5) $8! - 4! + 3 =$

6) $3! \times 4 - 12 =$

7) $(3! + 1!)^2 =$

8) $(5! - 4!)^2 =$

9) $(3!\,0!)^2 - 2 =$

10) $\dfrac{8!}{6!} =$

11) $\dfrac{3!}{2!} =$

12) $\dfrac{6!}{5!} =$

13) $\dfrac{21!}{19!} =$

14) $\dfrac{(n-1)!}{(n-3)!} =$

15) $\dfrac{(n+2)!}{(n+1)!} =$

16) $\dfrac{(4+2!)^3}{2!} =$

17) $\dfrac{4n!}{2n!} =$

18) $\dfrac{31!}{29!2!} =$

19) $\dfrac{13!}{9!3!} =$

20) $\dfrac{6 \times 280!}{3(4 \times 70)!} =$

21) $\dfrac{30!}{31!2!} =$

22) $\dfrac{7!7!}{8!5!} =$

23) $\dfrac{12!11!}{9!10!} =$

24) $\dfrac{(2 \times 5)!}{1!9!} =$

25) $\dfrac{2!(6n-1)!}{(6n)!} =$

26) $\dfrac{n(4n+4)!}{(4n+5)!} =$

27) $\dfrac{(n+1)!(n)}{(n+2)!} =$

Combinations and Permutations

✏ **Calculate the value of each.**

1) 6! = ____

2) 2! × 5! = ____

3) 4! = ____

4) 3! + 5! = ____

5) 7! = ____

6) 9! = ____

7) 3! + 3! = ____

8) 5! − 2! = ____

✏ **Find the answer for each word problems.**

9) Susan is baking cookies. She uses sugar, Vanilla and eggs. How many different orders of ingredients can she try? ____

10) Albert is planning for his vacation. He wants to go to museum, watch a movie, go to the beach, play volleyball and play football. How many ways of ordering are there for him? ____

11) How many 6-digit numbers can be named using the digits 1, 6, 8, 9, and 10 without repetition? ____

12) In how many ways can 4 boys be arranged in a straight line? ____

13) In how many ways can 8 athletes be arranged in a straight line? ____

14) A professor is going to arrange her 5 students in a straight line. In how many ways can she do this? ____

15) How many code symbols can be formed with the letters for the word FRIEND? ____

16) In how many ways a team of 7 basketball players can to choose a captain and co-captain? ____

Algebra 2 Workbook

Answers of Worksheets – Chapter 12

Mean and Median

1) Mean: 8.6, Median: 8
2) Mean: 19.43, Median: 17
3) Mean: 18.4, Median: 15
4) Mean: 17, Median: 12.5
5) Mean: 22, Median: 16
6) Mean: 93.5, Median: 49.5
7) Mean: 47, Median: 43.5
8) Mean: 21, Median: 17
9) Mean: 15.5, Median: 9
10) Mean: 22.88, Median: 19
11) Mean: 41.83, Median: 39
12) Mean: 25.25, Median: 14.5
13) Mean: 39.2, Median: 45
14) Mean: 6.6, Median: 7
15) 13

Mode and Range

1) Mode: 9, Range: 15
2) Mode: 8, Range: 13
3) Mode: 4, Range: 27
4) Mode: 10, Range: 16
5) Mode: 3, Range: 8
6) Mode: 7, Range: 22
7) Mode: 24, Range: 34
8) Mode: 13, Range: 11
9) Mode: 8, Range: 14
10) Mode: 14, Range: 28
11) Mode: 8, Range: 12
12) Mode: 12, Range: 28
13) Mode: 21, Range: 21
14) Mode: 10, Range: 25
15) 10

Probability Problems

1) $\frac{2}{5}$
2) $\frac{3}{11}, \frac{2}{5}$
3) $\frac{1}{5}$
4) $\frac{1}{13}$
5) $\frac{1}{5}$
6) $\frac{1}{2}$
7) $\frac{1}{2}$
8) $\frac{1}{8}$
9) 0

Factorials

1) 721
2) 122
3) 576
4) 714
5) 40,299
6) 12
7) 49
8) 9,216
9) 34
10) 56
11) 3
12) 6
13) 420
14) $(n-1)(n-2)$
15) $n+2$
16) 108
17) 2
18) 465
19) 2,860
20) 2
21) $\frac{1}{62}$
22) 5.25
23) 14,520
24) 10
25) $\frac{1}{3n}$
26) $\frac{n}{4n+5}$
27) $\frac{n}{n+2}$

Combinations and Permutations

1) 720
2) 240
3) 24
4) 126
5) 5,040
6) 362,880

WWW.MathNotion.Com

Algebra 2 Workbook

7) 12
8) 118
9) 6
10) 120

11) 720
12) 24
13) 40,320
14) 120

15) 720
16) 42

Algebra 2 Practice Tests

Time to Test

Time to refine your skill with a practice examination.

Take a REAL Algebra 2 test to simulate the test day experience. After you've finished, score your test using the answer key.

Before You Start

- You'll need a pencil, calculator, and a timer to take the test.
- It's okay to guess. You won't lose any points if you're wrong.
- After you've finished the test, review the answer key to see where you went wrong.

Graphing calculators are permitted for Algebra 2 Tests.

Good Luck!

Algebra Practice Tests Answer Sheet

Remove (photocopy) this answer sheet and use it to complete the practice test.

Algebra 1 Practice Test Answer Sheet

1 Ⓐ Ⓑ Ⓒ Ⓓ 11 Ⓐ Ⓑ Ⓒ Ⓓ 21 Ⓐ Ⓑ Ⓒ Ⓓ

2 Ⓐ Ⓑ Ⓒ Ⓓ 12 Ⓐ Ⓑ Ⓒ Ⓓ 22 Ⓐ Ⓑ Ⓒ Ⓓ

3 Ⓐ Ⓑ Ⓒ Ⓓ 13 Ⓐ Ⓑ Ⓒ Ⓓ 23 Ⓐ Ⓑ Ⓒ Ⓓ

4 Ⓐ Ⓑ Ⓒ Ⓓ 14 Ⓐ Ⓑ Ⓒ Ⓓ 24 Ⓐ Ⓑ Ⓒ Ⓓ

5 Ⓐ Ⓑ Ⓒ Ⓓ 15 Ⓐ Ⓑ Ⓒ Ⓓ 25 Ⓐ Ⓑ Ⓒ Ⓓ

6 Ⓐ Ⓑ Ⓒ Ⓓ 16 Ⓐ Ⓑ Ⓒ Ⓓ 26 Ⓐ Ⓑ Ⓒ Ⓓ

7 Ⓐ Ⓑ Ⓒ Ⓓ 17 Ⓐ Ⓑ Ⓒ Ⓓ 27 Ⓐ Ⓑ Ⓒ Ⓓ

8 Ⓐ Ⓑ Ⓒ Ⓓ 18 Ⓐ Ⓑ Ⓒ Ⓓ 28 Ⓐ Ⓑ Ⓒ Ⓓ

9 Ⓐ Ⓑ Ⓒ Ⓓ 19 Ⓐ Ⓑ Ⓒ Ⓓ 29 Ⓐ Ⓑ Ⓒ Ⓓ

10 Ⓐ Ⓑ Ⓒ Ⓓ 20 Ⓐ Ⓑ Ⓒ Ⓓ 30 Ⓐ Ⓑ Ⓒ Ⓓ

Algebra 2 Practice Test 1

✓ **30 Questions**

✓ **You may use a calculator for this test.**

Administered *Month Year*

Algebra 2 Workbook

1) If $\frac{5x}{36} = \frac{x-2}{6}$, $x = ?$

   A. $\frac{1}{12}$

   B. $\frac{3}{5}$

   C. 6

   D. 12

2) Five years ago, Amy was three times as old as Mike was. If Mike is 11 years old now, how old is Amy?

   A. 23

   B. 25

   C. 16

   D. 21

3) If $y = 2ab + 5b^2$, what is y when $a = 4$ and $b = 2$?

   A. 26

   B. 36

   C. 20

   D. 18

4) If $f(x) = 2 + 3x$ and $g(x) = -2x^2 - 6 - x$, then find $(g - f)(x)$?

   A. $2x^2 - 4x - 8$

   B. $2x^2 - 4x + 8$

   C. $-2x^2 - 4x + 8$

   D. $-2x^2 - 4x - 8$

5) What is the area of a square whose diagonal is 6 cm?

   A. 36 cm^2

   B. 18 cm^2

   C. 24 cm^2

   D. 12 cm^2

6) If $(x - 5)^2 = 9$ which of the following could be the value of $(x - 6)(x - 5)$?

   A. 2

   B. 6

   C. −2

   D. −5

WWW.MathNotion.Com

7) What is the value of x in the following figure?

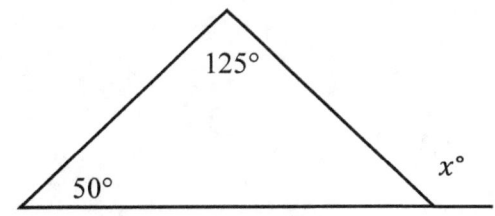

A. 125

B. 175

C. 55

D. 155

8) Right triangle ABC is shown below. Which of the following is true for all possible values of angle A and B?

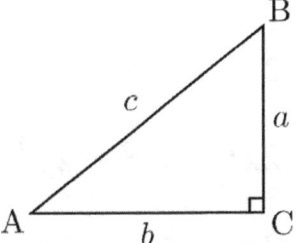

A. $\cot A = \cot B$

B. $\tan^2 A = \tan^2 B$

C. $\cos A = \sin B$

D. $\cot A = \cos B$

9) From the figure, which of the following must be true? (figure not drawn to scale)

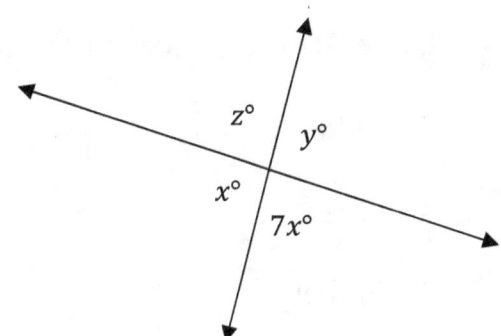

A. $y = 7x$

B. $y \geq x$

C. $y + 6x = z$

D. $z > x$

10) From last year, the price of gasoline has increased from $2.04 per gallon to $3.06 per gallon. The new price is what percent of the original price?

A. 52%

B. 150%

C. 130%

D. 170%

11) What is the value of x in the following system of equations?

$$x + 2y = 7$$
$$4x + 5y = 22$$

A. 3

B. 2

C. -3

D. 5

12) Simplify.

$$4x^2 + 5y^5 - 2x^2 + 5z^3 - y^2 + 4x^3 - 2y^5 + 3z^3$$

A. $2x^2 - 4y^2 + 3y^5 + 8z^3$

B. $2x^2 + 4x^3 - y^2 + 3y^5 + 8z^3$

C. $2x^2 - 4x^3 - 2y^2 + y^5 + 8z^3$

D. $2x^2 + 4x^3 - 3y^2 + 8z^3$

13) A ladder leans against a wall forming a 60° angle between the ground and the ladder. If the bottom of the ladder is 30 feet away from the wall, how long is the ladder?

A. 40 feet

B. 80 feet

C. 50 feet

D. 60 feet

14) The length of a rectangle is 5 meters greater than 7 times its width. The perimeter of the rectangle is 90 meters. What is the area of the rectangle?

A. 45 m²

B. 300 m²

C. 200 m²

D. 90 m²

Algebra 2 Workbook

15) Simplify $(-2 + 3i)(5 + 6i)$,

    A. $28 - 3i$

    B. $8 - 3i$

    C. $-10 + 3i$

    D. $-28 + 3i$

16) If $\tan \theta = \frac{5}{12}$ and $\sin \theta > 0$, then $\cos \theta = ?$

    A. $-\frac{5}{12}$

    B. $\frac{12}{13}$

    C. $\frac{5}{13}$

    D. $-\frac{13}{5}$

17) Which of the following has the half period and five times the amplitude of graph $y = \sin x$?

    A. $y = \frac{1}{2} \sin 5x$

    B. $y = 5\sin\left(\frac{x}{2} + 5\right)$

    C. $y = 5 + 5 \sin 2x$

    D. $y = 4 + \sin \frac{x}{2}$

18) y is $x\%$ of what number?

    A. $\frac{y}{100x}$

    B. $\frac{x}{100y}$

    C. $\frac{100y}{x}$

    D. $\frac{100x}{y}$

19) What is the solution of the following inequality?

$$|x - 5| \leq 2$$

    A. $x \geq 7 \cup x \leq 3$

    B. $3 \leq x \leq 7$

    C. $x \geq 10$

    D. $x \leq 7$

20) If cotangent of an angel β is $\sqrt{2}$, then the tangent of angle β is

A. $\frac{\sqrt{2}}{2}$

B. $-\frac{\sqrt{2}}{2}$

C. -1

D. 1

21) Which of the following points lies on the line $2x - 3y = 5$?

A. $(1, -2)$

B. $(-3, 0)$

C. $(-2, -3)$

D. $(1, -4)$

22) In the xy-plane, the point $(3, 9)$ and $(2, 8)$ are on line A. Which of the following equations of lines is parallel to line A?

A. $y = 2x$

B. $y = \frac{x}{3}$

C. $y = -x$

D. $y = x$

23) When point A $(7, 3)$ is reflected over the y-axis to get the point B, what are the coordinates of point B?

A. $(7, 3)$

B. $(-7, -3)$

C. $(-7, 3)$

D. $(7, -3)$

24) A bag contains 18 balls: three green, four black, six blue, a brown, two red and two white. If 17 balls are removed from the bag at random, what is the probability that a brown ball has been removed?

A. $\frac{1}{2}$

B. $\frac{1}{8}$

C. $\frac{17}{18}$

D. $\frac{1}{18}$

25) If 60% of x equal to 30% of 30, then what is the value of $(x + 3)^2$?

   A. 18.20

   B. 32.04

   C. 3,240

   D. 324

26) If $f(x) = 2x^3 + 3x^2 + x$ and $g(x) = -2$, what is the value of $f(g(x))$?

   A. 6

   B. 2

   C. 16

   D. -6

27) If $x \begin{bmatrix} 3 & 0 \\ 0 & 4 \end{bmatrix} = \begin{bmatrix} 2x + y - 5 & 0 \\ 0 & 3y - 12 \end{bmatrix}$, what is the product of x and y?

   A. 3

   B. 12

   C. 24

   D. 20

28) If $f(x) = 5^x$ and $g(x) = \log_5 x$, which of the following expressions is equal to $f(5g(p))$?

   A. $5P$

   B. 5^p

   C. p^5

   D. $\frac{p}{5}$

29) If one angle of a right triangle measures 30°, what is the sine of the other acute angle?

   A. $\frac{\sqrt{3}}{2}$

   B. $\frac{\sqrt{2}}{2}$

   C. $\frac{1}{2}$

   D. $\sqrt{2}$

30) In the following equation when z is divided by 6, what is the effect on x?

$$x = \frac{7y + \frac{r}{2r+3}}{\frac{12}{z}}$$

A. x is divided by 2

B. x is divided by 6

C. x is multiplied by 6

D. x is multiplied by 2

STOP

This is the End of this Test. You may check your work on this Test if you still have time.

Algebra 2 Practice Test 2

✓ **30 Questions**

✓ **You may use a calculator for this test.**

Administered *Month Year*

Algebra 2 Workbook

1) A number is chosen at random from 1 to 15. Find the probability of not selecting a composite number.

   A. $\dfrac{3}{15}$

   B. 15

   C. $\dfrac{2}{5}$

   D. 1

2) Simplify $\dfrac{3-4i}{-3i}$?

   A. $\dfrac{4}{3} + i$

   B. $\dfrac{4}{3} - i$

   C. $\dfrac{1}{3} - i$

   D. $\dfrac{1}{3} + i$

3) If $\sqrt{7x} = \sqrt{y}$, then $x =$?

   A. $\sqrt{\dfrac{y}{7}}$

   B. y^2

   C. $\sqrt{7y}$

   D. $\dfrac{y}{7}$

4) If $f(x) = 3x - 8$ and $g(x) = 2x^2 - 5x$, then find $\left(\dfrac{f}{g}\right)(x)$.

   A. $\dfrac{3x-8}{2x^2-5x}$

   B. $\dfrac{x-8}{2x^2-5x}$

   C. $\dfrac{x-5}{x^2-4}$

   D. $\dfrac{3x+8}{x^2+5x}$

5) In the standard (x, y) coordinate plane, which of the following lines contains the points $(1, -7)$ and $(6, 8)$?

   A. $y = 3x - 10$

   B. $y = -3x + 7$

   C. $y = -\dfrac{1}{3}x + 10$

   D. $y = 3x - 7$

6) If the interior angles of a quadrilateral are in the ratio 1:3:7:9, what is the measure of the largest angle?

  A. 54°   C. 126°

  B. 18°   D. 162°

7) If $x + 2sin^2 a + 2cos^2 a = 6$, then $x = ?$

  A. 2   C. 3

  B. 4   D. 6

8) An angle is equal to one fourth of its supplement. What is the measure of that angle?

  A. 36   C. 30

  B. 26.5   D. 60

9) If $sin\alpha = \frac{\sqrt{3}}{2}$ in a right triangle and the angle α is an acute angle, then what is $cos\,\alpha$?

  A. $\frac{\sqrt{3}}{3}$ C. $\sqrt{3}$

  B. $\frac{1}{3}$ D. $\frac{1}{2}$

10) What are the zeroes of the function $f(x) = 2x^3 + 14x^2 + 20x$?

  A. $-2, 4$ C. $0, -2, -5$

  B. $0, 2, 4$ D. $-2, -3$

11) In the standard (x, y) coordinate system plane, what is the area of the circle with the following equation?

$$(x + 2)^2 + (y - 1)^2 = 9$$

A. 3π

B. 9π

C. 18π

D. 6π

12) Simplify.

$$8x^5y^2 + 3x^3y^4 - (2x^5y^2 - 4x^3y^4)$$

A. $-x^5y^3$

B. $6x^5y^3 - 7x^5y^5$

C. $8x^5y^6$

D. $6x^5y^2 + 7x^3y^4$

13) In the following figure, what is the perimeter of $\triangle ABC$ if the area of $\triangle ADC$ is 45?

A. 57.5

B. 26

C. 25

D. 60

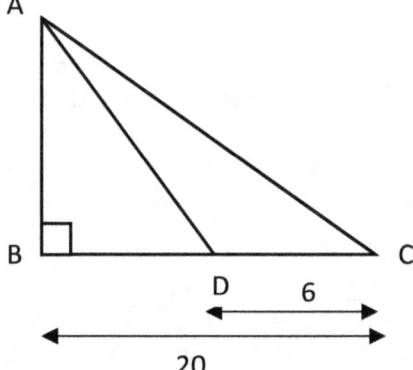

14) Which of the following is one solution of this equation?

$$3x^2 + 5x - 8 = 0$$

A. $\sqrt{5} + 1$

B. $\sqrt{5} - 1$

C. 1

D. $\sqrt{15}$

Algebra 2 Workbook

15) Three-kilograms apple and four-kilograms orange cost $51.2. If one-kilogram apple costs $2.4 how much does one-kilogram orange cost?

    A. $11

    B. $8

    C. $6.5

    D. $12

16) Which of the following expressions is equal to $\sqrt{\frac{3x^2}{5} + \frac{x^2}{25}}$?

    A. $4x$

    B. $\frac{4x}{5}$

    C. $2x\sqrt{x}$

    D. $\frac{x\sqrt{x}}{5}$

17) Tickets to a movie cost $10.50 for adults and $5.50 for students. A group of 18 friends purchased tickets for $119. How many student tickets did they buy?

    A. 4

    B. 14

    C. 11

    D. 18

18) If $x = 4$, what is the value of y in the following equation? $5y = \frac{3x^2}{8} + 9$

    A. 3

    B. 15

    C. 65

    D. 12

19) Let r and p be constants. If $x^2 + 4x + r$ factors into $(x + 3)(x + p)$, the values of r and p respectively are?

    A. 3, 1

    B. 1, 3

    C. 2, 3

    D. 3, 2

Algebra 2 Workbook

20) If 140% of a number is 70, then what is 80% of that number?

   A. 35

   B. 60

   C. 40

   D. 70

21) The average of six consecutive numbers is 24. What is the smallest number?

   A. 25

   B. 30.5

   C. 21.5

   D. 15.5

22) In a coordinate plane, triangle ABC has coordinates: $(5, -1)$, $(-4, -2)$, and $(2, 4)$. If triangle ABC is reflected over the y-axis, what are the coordinates of the new image?

   A. $(5, -1), (-2, -4), (4, 2)$

   B. $(-2, 4), (-4, -2), (5, -1)$

   C. $(-5, -1), (4, -2), (-2, 4)$

   D. $(4, -2), (-2, 4), (5, -2)$

23) What is the slope of a line that is perpendicular to the line $8x - 2y = 16$?

   A. -4

   B. $-\frac{1}{4}$

   C. 6

   D. 8

24) If $f(x) = 3x^3 + 5$ and $g(x) = \frac{2}{x}$, what is the value of $f(g(x))$?

   A. $\frac{8}{5x^3+5}$

   B. $\frac{5}{x^3}$

   C. $\frac{1}{5x+5}$

   D. $\frac{32}{x^3} + 5$

25) What is the solution of the following inequality?

$$|x - 4| \geq 7$$

A. $x \geq 11 \cup x \leq -3$

C. $x \geq 11$

B. $-3 \leq x \leq 11$

D. $x \leq -3$

26) If $\tan x = \frac{15}{20}$, then $\sin x =$

A. $\frac{1}{5}$

C. $\frac{12}{25}$

B. $\frac{15}{25}$

D. $\frac{7}{25}$

27) In the following figure, ABCD is a rectangle. If $a = \sqrt{2}$, and $b = 2a$, find the area of the shaded region. (the shaded region is a trapezoid)

A. 15

B. $12\sqrt{2}$

C. $8\sqrt{2}$

D. $4\sqrt{2}$

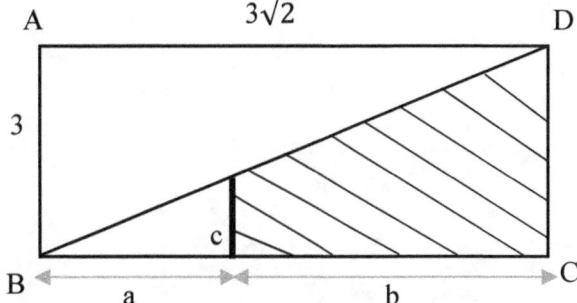

28) If the ratio of $9a$ to $8b$ is $\frac{1}{16}$, what is the ratio of a to b?

A. 16

C. $\frac{1}{18}$

B. 18

D. $\frac{1}{16}$

29) If $A = \begin{bmatrix} 1 & 1 \\ 2 & -1 \end{bmatrix}$ and $B = \begin{bmatrix} 4 & 2 \\ -2 & 3 \end{bmatrix}$, then $3A - B =$

A. $\begin{bmatrix} -3 & 1 \\ 4 & -4 \end{bmatrix}$

B. $\begin{bmatrix} -7 & 1 \\ 2 & -6 \end{bmatrix}$

C. $\begin{bmatrix} 4 & 2 \\ -1 & 3 \end{bmatrix}$

D. $\begin{bmatrix} -7 & 1 \\ 8 & -6 \end{bmatrix}$

30) What is the amplitude of the graph of the equation $y - 2 = 5\cos 2x$? (half the distance between the graph's minimum and maximum y-values in standard (x, y) coordinate plane is the amplitude of a graph.)

A. 2

B. 5

C. 4

D. 2.5

STOP

This is the End of this Test. You may check your work on this Test if you still have time.

Answers and Explanations

Algebra 2 Practice Tests

Answer Key

✳ Now, it is time to review your results to see where you went wrong and what areas you need to improve!

Algebra 2 Tests

Practice Test - 1

1	D	11	A	21	C
2	A	12	B	22	D
3	B	13	D	23	C
4	D	14	C	24	D
5	B	15	D	25	D
6	B	16	B	26	D
7	B	17	C	27	C
8	D	18	C	28	C
9	C	19	B	29	A
10	B	20	D	30	B

Practice Test - 2

1	C	11	B	21	C
2	A	12	D	22	C
3	D	13	D	23	B
4	A	14	C	24	D
5	A	15	A	25	A
6	D	16	B	26	B
7	B	17	B	27	D
8	A	18	A	28	C
9	D	19	A	29	D
10	C	20	C	30	B

Practice Tests 1
Answers and Explanations

1) Answer: D.

Solve for x, $\frac{5x}{36} = \frac{x-2}{6}$

Multiply the second fraction by 6, $\frac{5x}{36} = \frac{6(x-2)}{6 \times 6}$

Tow denominators are equal. Therefore, the numerators must be equal.

$5x = 6x - 12 \rightarrow -x = -12 \rightarrow x = 12$

2) Answer: A.

five years ago, Amy was three times as old as Mike. Mike is 11 years now. Therefore, 5 years ago Mike was 6 years.

five years ago, Amy was: $A = 3 \times 6 = 18$

Now Amy is 23 years old: $18 + 5 = 23$

3) Answer: B.

$y = 2ab + 5b^2$

Plug in the values of a and b in the equation: $a = 4$ and $b = 2$

$y = 2(4)(2) + 5(2)^2 = 16 + 5(4) = 16 + 20 = 36$

4) Answer: D.

$(g-f)(x) = g(x) - f(x) = (-2x^2 - 6 - x) - (2 + 3x)$

$-2x^2 - 6 - x - 2 - 3x = -2x^2 - 4x - 8$

5) Answer: B.

The diagonal of the square is 6. Let x be the side.

Use Pythagorean Theorem: $a^2 + b^2 = c^2$

$x^2 + x^2 = 6^2 \Rightarrow 2x^2 = 36 \Rightarrow 2x^2 = 36$

$\Rightarrow x^2 = 18 \Rightarrow x = \sqrt{18}$

The area of the square is: $\sqrt{18} \times \sqrt{18} = 18$

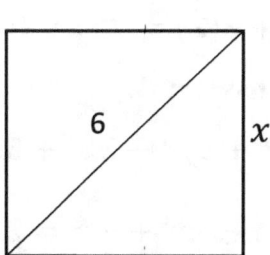

6) Answer: B.

$(x-5)^2 = 9 \to x-5 = 3 \to x = 8$

$\to (x-6)(x-5) = (8-6)(8-5) = (2)(3) = 6$

7) Answer: B.

$x = 50 + 125 = 175$

8) Answer: D.

By definition, the sine of any acute angle is equal to the cosine of its complement.

Since, angle A and B are complementary angles, therefore:

$\cos A = \sin B$

9) Answer: C.

x and z are colinear. y and $6x$ are colinear. Therefore,

$x + z = y + 7x,\ subtract\ x\ from\ both\ sides,\ then,\ z = y + 6x$

10) Answer: B.

The question is this: 3.06 is what percent of 2.04?

Use percent formula: $part = \frac{percent}{100} \times whole$

$3.06 = \frac{percent}{100} \times 2.04 \Rightarrow 3.06 = \frac{percent \times 2.04}{100} \Rightarrow 306 = percent \times 2.04$

$\Rightarrow percent = \frac{306}{2.04} = 150$

11) Answer: A.

Solving Systems of Equations by Elimination

Multiply the first equation by (–4), then add it to the second equation.

$\begin{matrix} -4(x+2y=7) \\ 4x+5y=22 \end{matrix} \Rightarrow \begin{matrix} -4x-8y=-28 \\ 4x+5y=22 \end{matrix} \Rightarrow -3y = -6 \Rightarrow y = 2$

Plug in the value of y into one of the equations and solve for x.

$x + 2(2) = 7 \Rightarrow x + 4 = 7 \Rightarrow x = 7 - 4 \Rightarrow x = 3$

12) Answer: B.

$4x^2 + 5y^5 - 2x^2 + 5z^3 - y^2 + 4x^3 - 2y^5 + 3z^3 = 4x^2 - 2x^2 + 4x^3 - y^2 + 5y^5 -$

$2y^5 + 5z^3 + 3z^3 = 2x^2 + 4x^3 - y^2 + 3y^5 + 8z^3$

13) Answer: D.

The relationship among all sides of special right triangle $30°, 60°, 90°$ is provided in this triangle:

In this triangle, the opposite side of $30°$ angle is half of the hypotenuse. Draw the shape of this question.

The ladder is the hypotenuse.

Therefore, the ladder is 60 ft.

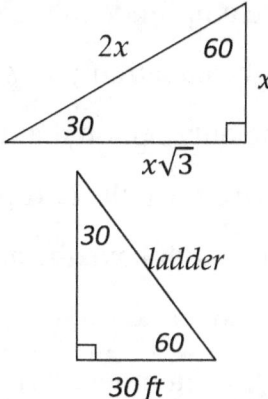

14) Answer: C.

Let L be the length of the rectangular and W be the with of the rectangular. Then, $L = 7W + 5$

The perimeter of the rectangle is 90 meters. Therefore:
$$2L + 2W = 90$$
$$L + W = 45$$

Replace the value of L from the first equation into the second equation and solve for W:
$$(7W + 5) + W = 45 \to 8W + 5 = 45 \to 8W = 40 \to W = 5$$

The width of the rectangle is 4 meters, and its length is:
$$L = 7W + 5 = 7(5) + 5 = 40$$

The area of the rectangle is: length × width = 40 × 5 = 200

15) Answer: D.

We know that: $i = \sqrt{-1} \Rightarrow i^2 = -1$
$$(-2 + 3i)(5 + 6i) = -10 - 12i + 15i + 18i^2 = -10 + 3i - 18 = 3i - 28$$

16) Answer: B.

$tan\theta = \frac{opposite}{adjacent}$

$tan\theta = \frac{5}{12} \Rightarrow$ we have the following right triangle. Then,

$c = \sqrt{5^2 + 12^2} = \sqrt{25 + 144} = \sqrt{169} = 13$

$cos\theta = \frac{adjacent}{hypotenuse} = \frac{12}{13}$

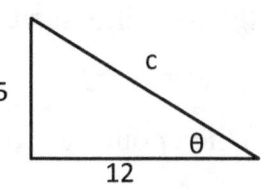

17) Answer: C.

The amplitude in the graph of the equation $y = a\sin bx$ is a. (a and b are constant)

In the equation $y = \sin x$, the amplitude is 1 and the period of the graph is 2π.

The only option that has five times the amplitude of graph $y = \sin x$ is $y = 5 + 5\sin 2x$ for the half period $\sin 2x = \sin 2\pi \Rightarrow 2x = 2\pi \Rightarrow x = \pi$

They both have the amplitude of 5 and period of π.

18) Answer: C.

Let the number be A. Then: $y = x\% \times A \rightarrow$ (Solve for A) $\rightarrow x = \dfrac{x}{100} \times A$

Multiply both sides by $\dfrac{100}{x}$: $y \times \dfrac{100}{x} = \dfrac{x}{100} \times \dfrac{100}{x} \times A \rightarrow A = \dfrac{100y}{x}$

19) Answer: B.

$|x - 5| \leq 2 \rightarrow -2 \leq x - 5 \leq 2 \rightarrow -2 + 5 \leq x - 5 + 5 \leq 2 + 5 \rightarrow 3 \leq x \leq 7$

20) Answer: D.

$tangent\ \beta = \dfrac{1}{cotangent\ \beta} = \dfrac{1}{\sqrt{2}} = \dfrac{\sqrt{2}}{2}$

21) Answer: C.

Plug in each pair of number in the equation:

A. $(1, -2)$: $2(1) - 3(-2) = 8$ Nope!

B. $(-3, 0)$: $2(-3) - 3(0) = -6$ Nope!

C. $(-2, -3)$: $2(-2) - 3(-3) = 5$ Bingo!

D. $(1, -4)$: $2(1) - 3(-4) = 14$ Nope!

E. $(0, -2)$: $2(0) - 3(-2) = 6$ Nope!

22) Answer: D.

The slop of line A is: $m = \dfrac{y_2 - y_1}{x_2 - x_1} = \dfrac{9 - 8}{3 - 2} = 1$

Parallel lines have the same slope and only choice E ($y = x$) has slope of 1.

23) Answer: C.

When points are reflected over y-axis, the value of y in the coordinates doesn't change and the sign of x changes. Therefore, the coordinates of point B is $(-7, 3)$.

Algebra 2 Workbook

24) Answer: D.

If 18 balls are removed from the bag at random, there will be one ball in the bag. The probability of choosing a brown ball is 1 out of 17. Therefore, the probability of not choosing a brown ball is 17 out of 18 and the probability of having not a brown ball after removing 17 balls is the same.

25) Answer: D.

$0.6x = (0.3) \times 30 \rightarrow x = 15 \rightarrow (x+3)^2 = (18)^2 = 324$

26) Answer: D.

$g(x) = -2$,

then $f(g(x)) = f(-2) = 2(-2)^3 + 3(-2)^2 + (-2) = -16 + 12 - 2 = -6$

27) Answer: C.

$\begin{cases} 3x = 2x + y - 5 \\ 4x = 3y - 12 \end{cases} \rightarrow \begin{cases} x - y = -5 \\ 4x - 3y = -12 \end{cases}$

Multiply first equation by -4.

$\begin{cases} -4x + 4y = 20 \\ 4x - 3y = -12 \end{cases} \rightarrow$ add two equations.

$y = 8 \rightarrow y = 8 \rightarrow x = 3 \rightarrow x \times y = 24$

28) Answer: C.

To solve for $f(5g(p))$, first, find $5g(p)$

$g(x) = \log_5 x \rightarrow g(p) = \log_5 p \rightarrow 5g(p) = 5\log_5 p = \log_5 p^5$

Now, find $f(5g(p))$: $f(x) = 5^x \rightarrow f(\log_5 p^5) = 5^{\log_5 p^5}$

Logarithms and exponentials with the same base cancel each other. This is true because logarithms and exponentials are inverse operations. Then: $f(\log_5 p^5) = 5^{\log_5 p^5} = p^5$

29) Answer: A.

The relationship among all sides of right triangle $30°, 60°, 90°$ is provided in the following triangle:

Sine of $60°$ equals to: $\dfrac{opposite}{hypotenuse} = \dfrac{x\sqrt{3}}{2x} = \dfrac{\sqrt{3}}{2}$

30) Answer: B.

$$x_1 = \frac{7y+\frac{r}{2r+3}}{\frac{z}{6}} = \frac{7y+\frac{r}{2r+3}}{\frac{6\times 12}{z}} = \frac{7y+\frac{r}{2r+3}}{6\times\frac{12}{z}} = \frac{1}{6}\times\frac{7y+\frac{r}{2r+3}}{\frac{12}{z}} = \frac{x}{6}$$

Practice Tests 2
Answers and Explanations

1) Answer: C.

Set of number that are not composite between 1 and 15: A = {2, 3, 5, 7, 11, 13}

Probability $= \frac{number\ of\ desired\ outcomes}{number\ of\ total\ outcomes} = \frac{6}{15} = \frac{2}{5}$

2) Answer: A.

To simplify the fraction, multiply both numerator and denominator by i.

$\frac{3-4i}{-3i} \times \frac{i}{i} = \frac{3i-4i^2}{-3i^2}$

$i^2 = -1$, Then: $\frac{3i-4i^2}{-3i^2} = \frac{3i-4(-1)}{-3(-1)} = \frac{3i+4}{3} = \frac{3i}{3} + \frac{4}{3} = i + \frac{4}{3}$

3) Answer: D.

Solve for x. $\sqrt{7x} = \sqrt{y}$

Square both sides of the equation: $(\sqrt{7x})^2 = (\sqrt{y})^2$

$7x = y \rightarrow x = \frac{y}{7}$

4) Answer: A.

$(\frac{f}{g})(x) = \frac{f(x)}{g(x)} = \frac{3x-8}{2x^2-5x}$

5) Answer: A.

The equation of a line is: $y = mx + b$, where m is the slope and b is the y-intercept.

First find the slope: $m = \frac{y_2-y_1}{x_2-x_1} = \frac{8-(-7)}{6-1} = \frac{15}{5} = 3$

Then, we have: $y = 3x + b$

Choose one point and plug in the values of x and y in the equation to solve for b.

Let's choose the point $(1, -7)$

$y = 3x + b \rightarrow -7 = 3(1) + b \rightarrow -7 = 3 + b \rightarrow b = -10$

The equation of the line is: $y = 3x - 10$

Algebra 2 Workbook

6) Answer: D.

The sum of all angles in a quadrilateral is 360 degrees.

Let x be the smallest angle in the quadrilateral. Then the angles are: $x, 3x, 7x, 9x$

$x + 3x + 7x + 9x = 360 \rightarrow 20x = 360 \rightarrow x = 18$

The angles in the quadrilateral are 18°, 54°, 126°, and 162°

7) Answer: B.

$2sin^2 a + 2cos^2 a = 2(sin^2 a + cos^2 a) = 2(1) = 2$, then:

$x + 2 = 6 \rightarrow x = 4$

8) Answer: A.

The sum of supplement angles is 180. Let x be that angle. Therefore, $x + 4x = 180 \Rightarrow$

$5x = 180$, divide both sides by 5: $x = 36$

9) Answer: D.

$sin\alpha = \frac{\sqrt{3}}{2} \Rightarrow$ Since $sin\alpha = \frac{opposite}{hypotenuse}$, we have the following right triangle. Then,

$c = \sqrt{2^2 - (\sqrt{3})^2} = \sqrt{4 - 3} = \sqrt{1} = 1$

$cos\alpha = \frac{1}{2}$

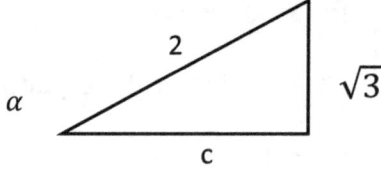

10) Answer: C.

Frist factor the function: $f(x) = 2x^3 + 14x^2 + 20x = 2x(x + 2)(x + 5)$

To find the zeros, $f(x)$ should be zero. $f(x) = 2x(x + 2)(x + 5) = 0$

Therefore, the zeros are: $x = 0$

$(x + 2) = 0 \Rightarrow x = -2$; $(x + 5) = 0 \Rightarrow x = -5$

11) Answer: B.

The equation of a circle in standard form is:

$(x - h)^2 + (y - k)^2 = r^2$, where r is the radius of the circle.

In this circle the radius is 3. $r^2 = 9 \rightarrow r = 3$

$(x + 2)^2 + (y - 1)^2 = 3^2$

Area of a circle: $A = \pi r^2 = \pi(3)^2 = 9\pi$

WWW.MathNotion.Com

12) Answer: D.

$8x^5y^2 + 3x^3y^4 - (2x^5y^2 - 4x^3y^4) = 8x^5y^2 - 2x^5y^2 + 3x^3y^4 + 4x^3y^4 = 6x^5y^2 + 7x^3y^4$

13) Answer: D.

Let x be the length of AB, then: $45 = \frac{x \times 6}{2} \rightarrow x = 15$

The length of $AC = \sqrt{15^2 + 20^2} = \sqrt{625} = 25$

The perimeter of $\Delta ABC = 15 + 20 + 25 = 60$

14) Answer: C.

$x_{1,2} = \frac{-b \pm \sqrt{b^2 - 4ac}}{2a}$

$ax^2 + bx + c = 0 \Rightarrow 3x^2 + 5x - 8 = 0$, then: a = 3, b = 5 and c = $-$ 8

$x = \frac{-5 + \sqrt{5^2 - 4 \times 3 \times (-8)}}{2 \times 3} = 1$; $x = \frac{-5 - \sqrt{5^2 - 4 \times 3 \times (-8)}}{2 \times 3} = -\frac{8}{3}$

15) Answer: A.

Let x be the cost of one-kilogram orange, then: $4x + (3 \times 2.4) = 51.2$

$\rightarrow 4x + 7.2 = 51.2 \rightarrow 4x = 51.2 - 7.2 \rightarrow 4x = 44 \rightarrow x = \frac{44}{4} = \$11$

16) Answer: B.

Simplify the expression.

$\sqrt{\frac{3x^2}{5} + \frac{x^2}{25}} = \sqrt{\frac{15x^2}{25} + \frac{x^2}{25}} = \sqrt{\frac{16x^2}{25}} = \sqrt{\frac{16}{25}x^2} = \sqrt{\frac{16}{25}} \times \sqrt{x^2} = \frac{4}{5} \times x = \frac{4x}{5}$

17) Answer: B.

Let x be the number of adult tickets and y be the number of student tickets. Then:

$x + y = 18$

$10.50x + 5.50y = 119$

Use elimination method to solve this system of equation. Multiply the first equation by -5.5 and add it to the second equation.

$-5.5(x + y = 18) \Rightarrow -5.5x - 5.5y = -99$

$10.50x + 5.50y = 119 \Rightarrow 5x = 20 \rightarrow x = 4$

There are 4 adults' tickets and 14 student tickets.

18) Answer: A.

Plug in the value of x in the equation and solve for y.

$5y = \frac{3x^2}{8} + 9 \to 5y = \frac{3(4)^2}{8} + 9 \to 5y = \frac{3(16)}{8} + 9 \to 5y = 6 + 9 = 15$

$\to 5y = 15 \to y = 3$

19) Answer: A.

$(x + 3)(x + p) = x^2 + (3 + p)x + 3p \to 3 + p = 4 \to p = 1$ and $r = 3p = 3$

20) Answer: C.

First, find the number.

Let x be the number. Write the equation and solve for x.

140% of a number is 70, then:

$1.4 \times x = 70 \Rightarrow x = 70 \div 1.4 = 50$

80% of 50 is: $0.8 \times 50 = 40$

21) Answer: C.

Let x be the smallest number. Then, these are the numbers:

$x, x + 1, x + 2, x + 3, x + 4, x + 5$

$\text{average} = \frac{\text{sum of terms}}{\text{number of terms}} \Rightarrow 24 = \frac{x+(x+1)+(x+2)+(x+3)+(x+4)+(x+5)}{5} \Rightarrow 24 = \frac{6x+15}{6} \Rightarrow 144$

$= 6x + 15 \Rightarrow 129 = 6x \Rightarrow x = 21.5$

22) Answer: C.

Since the triangle ABC is reflected over the y-axis, then all values of y's of the points don't change and the sign of all x's change. (remember that when a point is reflected over the y-axis, the value of y does not change and when a point is reflected over the x-axis, the value of x does not change). Therefore:

$(5, -1)$ changes to $(-5, -1)$

$(-4, -2)$ changes to $(4, -2)$

$(2, 4)$ changes to $(-2, 4)$

23) Answer: B.

The equation of a line in slope intercept form is: $y = mx + b$

Solve for y.

$8x - 2y = 16 \Rightarrow -2y = 16 - 8x \Rightarrow y = (16 - 8x) \div (-2) \Rightarrow$

$y = 4x - 8 \rightarrow$ The slope is 4.

The slope of the line perpendicular to this line is:

$m_1 \times m_2 = -1 \Rightarrow 4 \times m_2 = -1 \Rightarrow m_2 = -\frac{1}{4}$

24) Answer: D.

$f(g(x)) = 4 \times \left(\frac{2}{x}\right)^3 + 5 = \frac{32}{x^3} + 5$

25) Answer: A.

$x - 4 \geq 7 \rightarrow x \geq 7 + 4 \rightarrow x \geq 11$

Or $x - 4 \leq -7 \rightarrow x \leq -7 + 4 \rightarrow x \leq -3$

Then, solution is: $x \geq 11 \cup x \leq -3$

26) Answer: B.

$\tan = \frac{opposite}{adjacent}$, and $\tan x = \frac{15}{20}$, therefore, the opposite side of the angle x is 15 and the adjacent side is 20. Let's draw the triangle.

Using Pythagorean theorem, we have:

$a^2 + b^2 = c^2 \rightarrow 15^2 + 20^2 = c^2 \rightarrow 225 + 400 = c^2 \rightarrow c = 25$

$\sin x = \frac{opposite}{hypotenuse} = \frac{15}{25} = \frac{3}{5}$

27) Answer: D.

Based on triangle similarity theorem:

$\frac{a}{a+b} = \frac{c}{3} \rightarrow c = \frac{3a}{a+b} = \frac{3\sqrt{2}}{\sqrt{2} + 2\sqrt{2}} = 1$

$\rightarrow$ area of shaded region is: $\left(\frac{c+3}{2}\right)(b) = \frac{4}{2} \times 2\sqrt{2} = 4\sqrt{2}$

28) Answer: C.

Write the ratio of $9a$ to $8b$, $\frac{9a}{8b} = \frac{1}{16}$

Use cross multiplication and then simplify.

$9a \times 16 = 8b \times 1 \rightarrow 144a = 8b \rightarrow a = \frac{8b}{144} = \frac{b}{18}$

Now, find the ratio of a to b. $\frac{a}{b} = \frac{\frac{b}{18}}{b} \rightarrow \frac{b}{18} \div b = \frac{b}{18} \times \frac{1}{b} = \frac{b}{18b} = \frac{1}{18}$

29) Answer: D.

First, find $3A$.

$$A = \begin{bmatrix} 1 & 1 \\ 2 & -1 \end{bmatrix} \Rightarrow 3A = 3 \times \begin{bmatrix} 1 & 1 \\ 2 & -1 \end{bmatrix} = \begin{bmatrix} 3 & 3 \\ 6 & -3 \end{bmatrix}$$

Now, solve for $3A - B$:

$$\begin{bmatrix} -3 & 3 \\ 6 & -3 \end{bmatrix} - \begin{bmatrix} 4 & 2 \\ -2 & 3 \end{bmatrix} = \begin{bmatrix} -3-4 & 3-2 \\ 6-(-2) & -3-3 \end{bmatrix} = \begin{bmatrix} -7 & 1 \\ 8 & -6 \end{bmatrix}$$

30) Answer: B.

The amplitude in the graph of the equation $y = a\cos bx$ is a. (a and b are constant)

In the equation $y - 2 = 5\cos 2x$, the amplitude is 5.

"End"

www.ingramcontent.com/pod-product-compliance
Lightning Source LLC
Chambersburg PA
CBHW081110080526
44587CB00021B/3531